ANATOMY
AND THE
PROBLEM OF BEHAVIOUR

Cambridge University Press
Fetter Lane, London
New York
Bombay, Calcutta, Madras
Toronto
Macmillan
Tokyo
Maruzen-Kabushiki-Kaisha

ANATOMY
AND THE
PROBLEM OF BEHAVIOUR

by

G. E. COGHILL

Member of the Wistar Institute
of Anatomy and Biology
Philadelphia

LECTURES DELIVERED AT
UNIVERSITY COLLEGE
LONDON

CAMBRIDGE
AT THE UNIVERSITY PRESS
1929

PRINTED IN GREAT BRITAIN

PREFACE

My earliest scientific interests were aroused by an introductory collegiate course in psychology under Professor E. B. Delabarre. This course of instruction left me in an inquiring state of mind concerning what seemed to me to be the fundamental principles of psychology—the nature and interrelation of sensation, perception and thought. Eventually this attitude ripened into a decision to enter upon graduate study in psychology. But before an opportunity came for me to carry out this decision I became aware that the natural approach to the kind of psychological information I wanted lay through the physiology of the nervous system. Obviously, also, the physiology of the nervous system must be approached through its anatomy, about which I knew nothing.

As a matter of great good fortune, just at this juncture in my reasoning, I fell under the immediate instruction and inspiration of the late Professor Clarence Luther Herrick, who, with his characteristic personal devotion to the student, and with all the material resources at his command, encouraged me to begin studies on the comparative anatomy of the nervous system, and suggested *Amblystoma* as an appropriate form for my investigations. Work of this nature, the purpose of which was to analyse the peripheral nervous system upon the basis of functional units according to the conception of nerve components, brought me at once under the influence of my first great teacher's brother, Professor C. Judson Herrick. In the first of these two brothers I found the inspiration and opportunity to enter the field of scientific inquiry; in the second, I have enjoyed these many years a constant influence of orientation and stimulation in neurological work. To these two men I cannot estimate how much I owe in the methods of study and interpretation of the questions presented in the lectures which constitute this book. Certainly my debt to them is very great.

In the progress of my studies I soon met the conventional neuroanatomy; but this was disappointing to me, as it apparently still is to those who are psychologically inclined, as a foundation for thinking along the lines of my chief interests. Certain fundamentals for physiology and pathology, of course, were there: the neurone, the conduction path, cortical localization of motor functions, sense organs, etc.; all leading up to the conception of the nervous system as a telephone or telegraph with its conductors carrying "messages," a conception that seemed to me utterly inadequate and misleading. Nevertheless, questions concerning the comparative anatomy of functional units of the nervous system held my interest as a possible approach to a more satisfactory interpretation of the nervous system until the rapidly developing literature upon animal behaviour, or comparative psychology, suggested to me the idea of making parallel studies of the development of behaviour and the development of the nervous system. It seemed to me basic to a scientific study of behaviour to know whether the behaviour pattern of an animal develops haphazard or in an orderly manner; and that, if it should be found that behaviour develops in an orderly manner, then there should be a corresponding order of development structurally and functionally in the nervous system. Should such correlation be found to occur, it seemed to me that the development of behaviour would offer a new basis for the interpretation of nervous structure and function, and that the development of the nervous system would throw new light on problems of behaviour. Meanwhile I found myself engaged in neurological studies upon animals that were ideal for such a correlated anatomical and physiological inquiry, *Amblystoma* and its near relatives.

At about the time I began on the Pacific coast my studies upon the correlation of structure and function in the development of the nervous system of amphibians, Stewart Paton began similar studies upon amphibians and fishes on the Atlantic coast, each of us unaware of the other's work or interests. Paton began his work, as he states, with the intention of describing "some of the more important of the earliest reactions of the embryo to stimulation, and then noting in a parallel column the synchronous morphological changes in the nervous system." The aim of his first published report upon his studies (The reaction of the vertebrate embryo to stimulation and the associated changes in the nervous system. Mittheilungen aus die zoologische Station zu Neapel, Bd. 18, June, 1907) was "to describe some of the more striking phenomena that occur in the embryo at the time when the first cardiac beats and earliest responses to external stimulation begin, and then to determine in a general, but not in a specific, way how far these reactions are dependent upon the functional activity of the nervous system." "This problem," he says, "is essentially different from the far more difficult task of trying to determine the links between the two sets of facts." His conclusions relate to the myogenic and neurogenic origin of movement and to the origin and physiological significance of neurofibrils. Paton's studies on the correlation of structure and function in the development of the nervous system are, therefore, along very different lines from mine. His correlations concern general processes and general structures; mine relate to specific relations in definite behaviour processes (see citation of literature, 12 *a*, *b*, *c*).

The leading results of my studies upon the correlation of structural development of the nervous system with the development of behaviour are presented in the first lecture. But the work is by no means complete. Although, from what has already been accomplished, it is clear how the nervous system maintains the integrity of the individual while the behaviour expands in scope and increases in complexity, it remains still to determine more precisely the structural basis for the individuation of partial patterns (reflexes) within the total pattern. Of course it is not anticipated that every response of adult animals can ever be exactly explained in detail by structural organization and development; but it is reasonable to hope that the general mechanistic and dynamic principles that are basic to the individuation and conditioning of behaviour can be determined, and that the anatomical foundation of these principles can be made of much more immediate use to psychology than the conventional anatomy that has long been in vogue as an introduction to elementary textbooks on the subject.

The topic treated in the second lecture thrust itself

into my problem in the early period of my studies, particularly in connection with the origin of the motor root fibres; and it was about this time that Professor C. M. Child began to publish his interpretation of the physiological gradient as a factor in the regulation and development of organisms. This interpretation appealed to me at once as possibly applicable to the problem of why conduction paths grow as they do in the central nervous system of *Amblystoma*, and soon after this Professor Child published his book, *The origin and development of the nervous system*, which makes specific application of the principle of physiological gradients to the development of the nervous system. Although, as pointed out in the text, there are minor differences in detail between Professor Child's interpretation and mine, I cannot make too generous acknowledgment to him for the use I have here made of his conception of gradients.

In the third lecture my original interest in psychology presents itself in an effort to interpret, in terms of behaviour, the results that are recorded in the first and second lectures. But I must state here definitely that I do not regard myself as a psychologist. My ambition to study psychology still lingers; but it can in all probability never be realized. If this attempt to apply new facts to the behaviour problem, or to psychology in its broadest sense, should prove to be wide of the mark, the facts themselves remain. It is hoped at least that their presentation in the present necessarily brief form may lead to wholesome discussion and eventually make for progress in biological interpretations of psychological problems.

I am indebted to Dr Engelbrekt A. Swenson for the motion pictures from which Figs. 3, 4 and 5 were traced; and to Dr R. C. Baker for permission to use Figs. 33 to 41 inclusive. The drawings are by Miss Dorothy Harris.

Especially would I gratefully acknowledge my indebtedness to the University of London for the privilege of speaking upon my work in University College, and to Professor G. Elliot Smith and his colleagues of the Institute of Anatomy for their gracious personal hospitality while I was a guest among them.

G. E. C.

THE WISTAR INSTITUTE OF ANATOMY AND BIOLOGY
PHILADELPHIA, PENNSYLVANIA, U.S.A.
15 *June* 1928

CONTENTS

LECTURE III

Growth of the Nerve Cell and the Interpretation of Behaviour

Lecture One

THE DEVELOPMENT OF BEHAVIOUR AND ITS ANATOMICAL EXPLANATION IN A TYPICAL VERTEBRATE

THE work upon which you have asked me to address you has to do with the development of behaviour and the growth of the nervous system. It had its inception in the relatively early days of the modern school of animal behaviour. My earlier investigations had been neurological: physiological in motive, but morphological in method; consisting chiefly of the analysis of the peripheral nervous system according to functional units (11, 12). That work had been exclusively upon the Amphibia, but chiefly upon *Amblystoma*. With this animal already in hand as a tool for neurological inquiry I undertook to analyse behaviour by studying its development, with the thought in mind that such an analysis should constitute a new approach to fundamental problems concerning the function of the nervous system and its parts, and that anatomical studies correlated throughout with the behaviour studies should throw new light upon the problem of behaviour. My later work, accordingly, has consisted of studies of the behaviour closely correlated at all phases with studies of the structural development of the nervous system of *Amblystoma*.

THE ADAPTABILITY OF *AMBLYSTOMA* TO THE PROBLEM

The adaptability of this animal to my purpose is an important consideration (Fig. 1). Its practically total lack of specialization in structure is obvious. The head is symmetrically formed and the organs of special sense, for vision, smell and taste, are well developed, but not excessively. The ear is ordinary as an organ of equilibration; as an organ of hearing it is very lowly developed. The flexible neck, trunk and tail merge into one symmetrical organ of locomotion of the most primitive

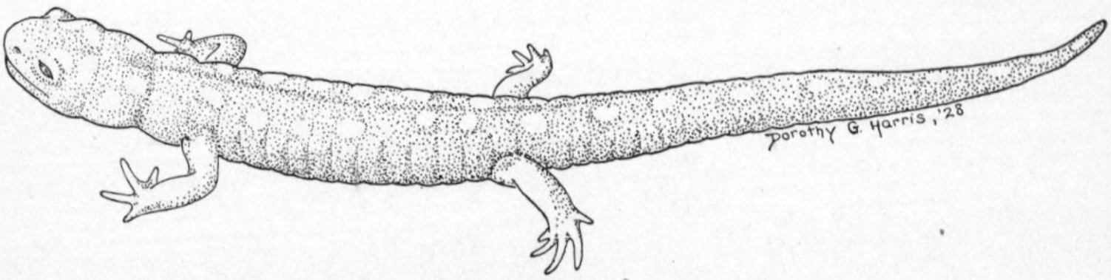

Fig. 1. Drawing of *Amblystoma punctatum*. Length of adult about 15 cm.

nature among vertebrates. The fore limbs and hind limbs are about equally developed. They are of moderate size, and digitate. There is even lack of noteworthy specialization among the digits. Finally, *Amblystoma* not only lacks structural specialization, but it lacks also evidence of retrogressive evolution in any important feature. Upon morphological grounds, therefore, our experimental animal must be regarded as relatively primitive, and typical among vertebrates.

Certain physiological features of *Amblystoma* are also of vital concern to this investigation. By reason of the equal division of the egg in the cleavage stages every cell of the embryo receives its own supply of food substance in the form of yolk. This food is adequate for the activities of all of the cells of the animal until after the neuromuscular system has reached a relatively advanced stage of functional development, for the animals do not take food until several days after they begin to swim. There is, therefore, in connection with our observations, no possible complication with a variable factor of food

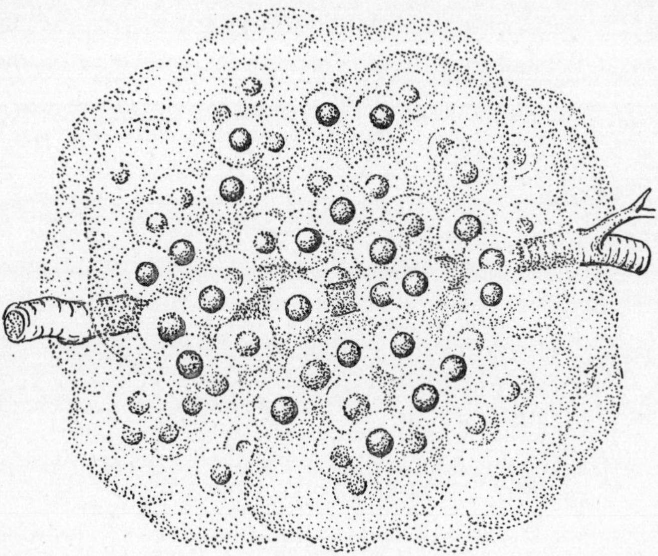

Fig. 2. Drawing of a clutch of eggs of *Amblystoma punctatum*. Diameter of the eggs about 2 mm.

supply. The egg-laying habit, also, is peculiarly advantageous, for eggs are deposited in large numbers in clutches, so that many embryos of the same age can be had for experimentation from time to time (Fig. 2). The embryos are readily removed from the membranes without injury; and they can be kept under observation continuously from the first muscular contraction to the

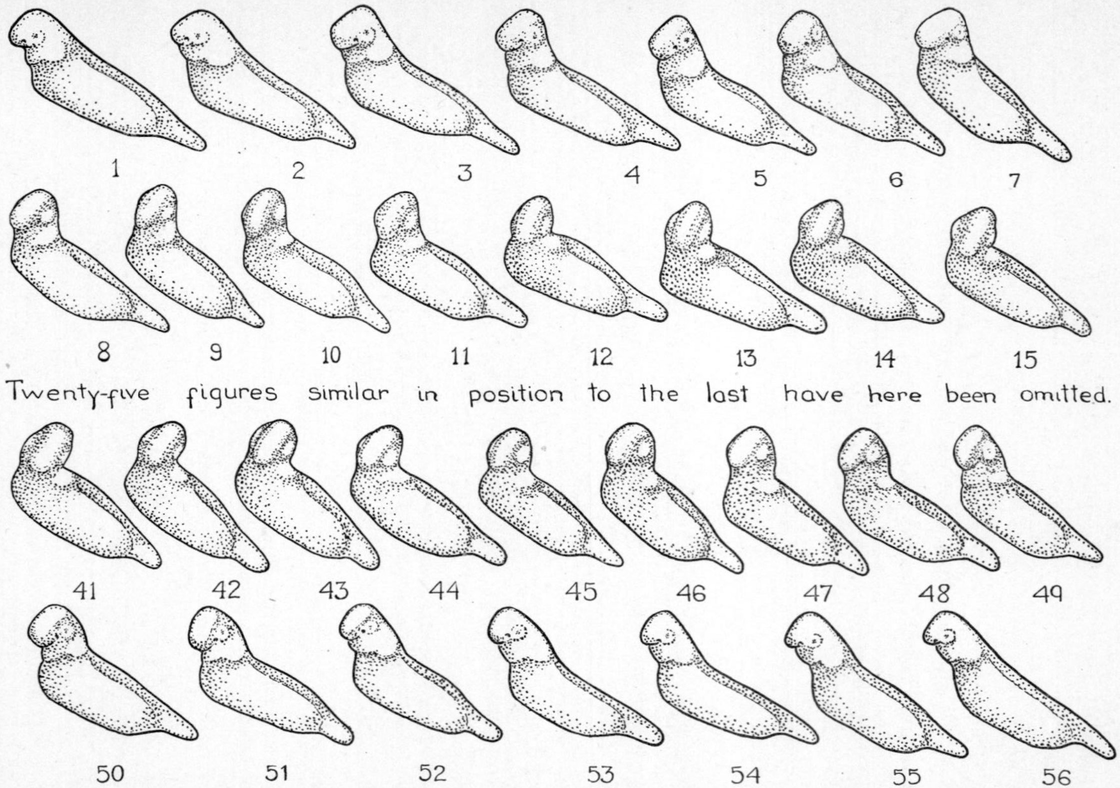

Fig. 3. Tracing of motion pictures of *Amblystoma tigrinum* (Axolotl) taken while performing the early flexure. Camera speed, 17 pictures per second. Twenty-five pictures are omitted from the middle of the scene. The sharp localization of the intensive contraction and its maintenance through a relatively long period are features of importance.

perfection of the pattern of behaviour. Indeed, there is probably no other animal that offers better advantages than *Amblystoma* presents for the search after general principles of behaviour and nervous function in the vertebrates.

It is the purpose of this lecture, first, to trace the course of development of behaviour in *Amblystoma*, and, second, to show how the nervous system plays its rôle in this development.

THE DEVELOPMENT OF AQUATIC LOCOMOTION

The first movement that is executed by *Amblystoma* is a bending of the head to one side (Fig. 3). This is performed by the contraction of the muscle segments situated immediately behind the head. This movement is slow in contraction and slow in relaxation. As the embryo advances in age the muscular contractions extend farther down the trunk until, in the course of about 36 hours, the entire trunk is involved in the performance (Fig. 4). The animal now bends itself into a tight coil. The coil may be to the right or to the left, or a coil in one direction may be reversed instantly into a coil in the opposite direction. The muscular contraction and relaxation have now acquired considerable speed.

Fig. 4. A tracing of one of the photographs of a motion picture series of the embryo of *Amblystoma tigrinum* (Axolotl) taken while performing the coil reaction. This coil was maintained for approximately a second and a half. The entire reaction lasted four seconds.

During this early period of behaviour all muscular contractions begin in the head region and progress caudad: the individual performance recapitulates the history of its development. The history has been a

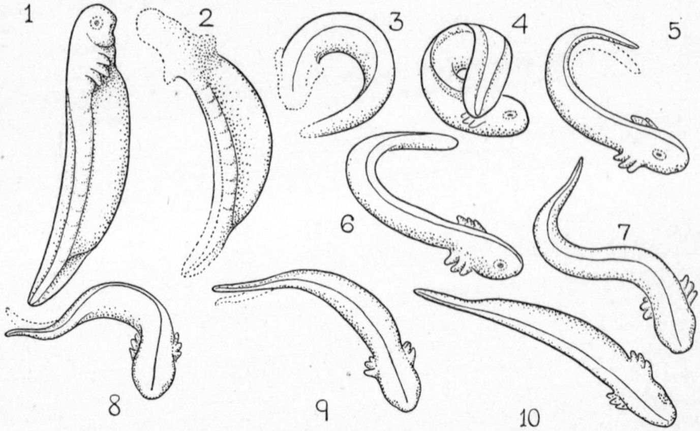

Fig. 5. Tracings (enlarged) of motion pictures of *Amblystoma tigrinum* (Axolotl) performing the early swimming movement. The animal is at rest in position 1. In 2 the head is moving to the left so rapidly that it blurs. Through 3 and 4 it completes a coil. In 5 and 6 the anterior part of the trunk has straightened out, but the original flexure still affects the caudal part. In 7 a flexure to the right has begun, while the original flexure is passing tailward. In 8 the first flexure has almost disappeared, while in 9 the animal is straightening out to its resting position in 10. The actual positions of the figures with reference to each other on the page have no significance; but the positions of the figures with reference to the points of the compass are correct. In passing from position 1 to position 10 the animal changes its orientation as indicated in the figures. The change of orientation (the direction in which the head points) is effected by the movements 1 to 5. From 5 to 10 the animal moves forward approximately its body length. This is determined by reference to a body that remained at rest in the field. Camera rate, 17 pictures per second. The reaction lasted approximately three-fifths of a second.

gradual development of the simple flexure into its greatest possible capacity of performance. Nothing really new has yet been introduced into the behaviour pattern of the animal since its first movement was per-

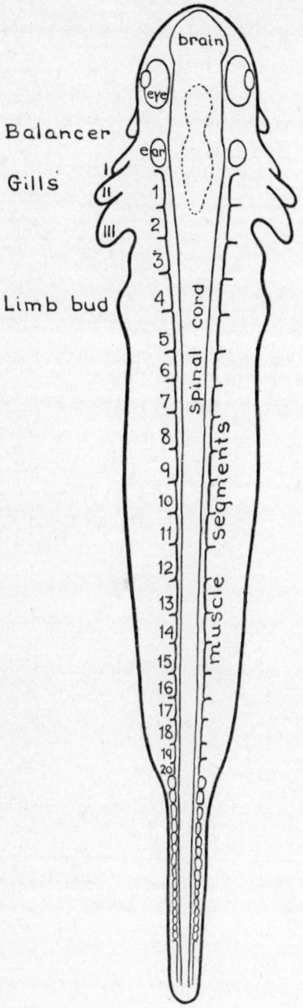

Fig. 6

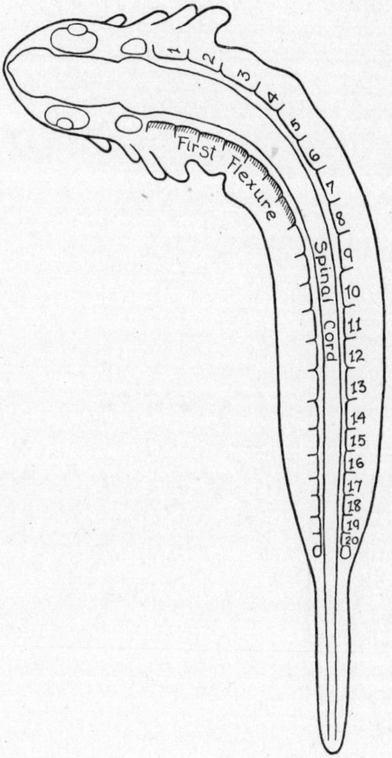

Fig. 7

Fig. 6. A diagram of *Amblystoma punctatum* of the early swimming stage, made to scale from serial sections to show important structures as seen from the dorsal side. The arrangement of the muscle segments alongside of the spinal cord and the absence of limbs excepting very early primordia of the fore limbs are important features. Actual length of the embryo approximately 7 mm. Modified from Fig. 4 of Herrick and Coghill (28).

Fig. 7. A diagram modified from that of Fig. 6 to illustrate the beginning of a swimming movement as a first flexure by contraction of a number of the anterior muscle segments, indicated by cross-hatching. Modified from Fig. 5 of Herrick and Coghill (28).

formed, and the coil reaction gives the animal no locomotor power. Nevertheless the coil has in it the primary locomotor factor: cephalocaudal progression of muscular contraction.

But from the coil the animal turns instantaneously into a new performance that solves its problem of locomotion (Fig. 5). A flexure begins in the head region and progresses caudad as if, according to its habit, the animal were going to throw its body into a tight coil; but instead of doing this it reverses the flexure in the anterior part before the first flexure has passed through the entire length of the animal. There are now two flexures in process at the same time, one to the right and the other to the left, and both of them progressing from the head tailward (Figs. 6, 7, 8). Meanwhile the movements are increasing individually in speed and serially in duration, and, through their progression from the head tailward, they exercise a pressure upon the water in such a way as to drive the animal forward. In this manner locomotion is accomplished.

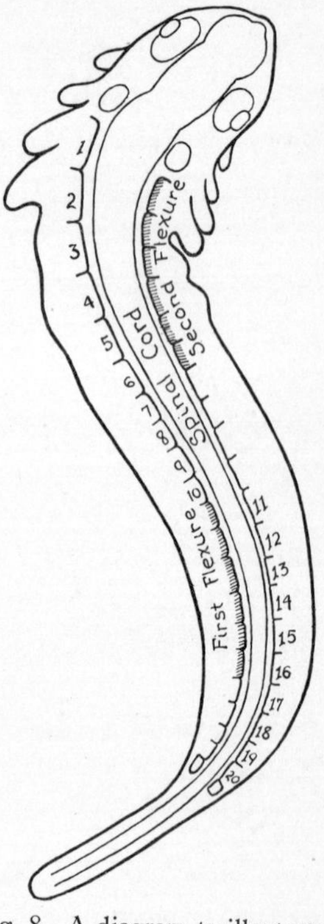

Fig. 8. A diagram to illustrate the swimming movement in which the first flexure has passed tailward and the second flexure is beginning in the anterior region by contraction of the muscle segments indicated by cross-hatching. Modified from Fig. 6 of Herrick and Coghill (28).

FIVE PHYSIOLOGICAL STAGES

This period of development falls into five stages which are designated as follows: (1) the non-motile stage, in which the muscles can be excited to contraction by direct stimulation, as by the stab of a sharp needle, by mechanical impact or by electricity, but cannot be excited by light touch on the skin; (2) the early flexure stage, when the animal first responds to light touch on the skin; (3) the coil stage, marked by the bending into a tight coil; (4) the "S" reaction, which is characterized by the reversal of a flexure before it is completely executed as a coil; and (5) the performance of the "S" reaction in series sufficient to effect locomotion. The first stage marks muscle contractility without nervous control; the second, earliest nervous excitation of the muscles from the tactile receptor field; the third, the full development of muscular strength; the fourth, the earliest co-ordination for locomotion; the fifth, locomotion accomplished.

With the attainment of locomotion *Amblystoma* has passed one of the most significant landmarks in the evolution of animal behaviour. It is no longer helplessly lodged in a fixed position and dependent upon fluctuations in the forces of nature around it for new experience with the environment; it can now explore new worlds through the action of internal forces in response to stimulation from without, or under the urge of its own dynamic mechanism.

How the mechanism arises as a dynamic system may be fully appreciated only by tracing its development through the preneural stages, but, leaving this history for consideration in the second lecture, we pass directly to the explanation of the behaviour in terms of functional nervous mechanisms.

ANATOMICAL EXPLANATION OF THE COIL

The muscular system at the beginning of motility consists of two longitudinal series of muscle segments,

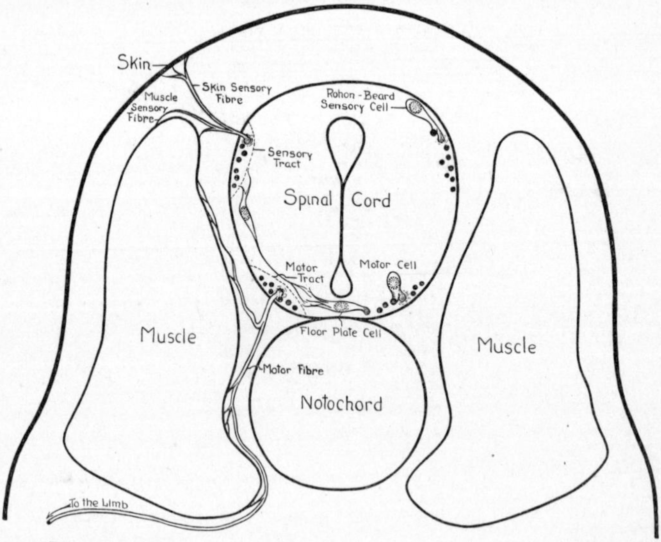

Fig. 9. Diagram of the relation of the nervous system of *Amblystoma* to the muscles and skin in the transverse plane of section. On the right side are the sensory cells dorsally and the motor cells ventrally. The processes of both sensory and motor cells branch into ascending and descending fibres, forming a sensory or afferent tract on the one hand and a motor or efferent tract on the other. On the left side of the figure the branching of a sensory fibre to innervate both skin and muscle is illustrated dorsally, and ventrally a motor fibre that innervates the muscle segment is shown with a branch going beyond the muscle segment to the primordium of the fore limb.

one on either side of the spinal cord and notochord, pressing closely upon these structures (Figs. 6, 9). Excepting the heart, these muscle segments or myotomes constitute the only functional muscle in *Amblystoma* until after it begins to swim. The behaviour just described, therefore, constitutes the entire somatic behaviour pattern of the animal at the age under consideration, and the muscle segments are the only effectors in it.

In the non-motile stage, that is, when the animal can contract its muscles but cannot do so in response to the stimulus of light touch upon the skin, there are motor nerve roots to about twelve of the most anterior muscle segments. These root fibres are naked protoplasmic threads that grow from nerve cells in the spinal cord to the middle of the muscle segment (18). The central cells from which they grow lie in the ventral tract of the cord and form a longitudinal path or tract which conducts impulses from the head tailward (Fig. 9).

In the dorsal part of the cord (Fig. 9) are cells structurally like these motor cells in that they form a longitudinal tract in the spinal cord and send branches out of it, but they are sensory in function instead of motor (13, 28). Their branches that grow out of the cord go to the skin and to the ends of the muscle segment, instead of to its central part (Fig. 10). Many if not all of the fibres of this system are both muscle-sensory (proprioceptive) and skin-sensory (exteroceptive). The neurones from which they arise constitute a longitudinal path conducting headward.

There are therefore in this non-motile *Amblystoma* both sensory and motor nerves in contact with their respective organs, the sensory field on the one hand and the muscles on the other, but the anatomical relations between these systems are such that an excitation cannot pass from the sensory to the motor mechanism. The muscles, while contracting in response to a mechanical stimulus applied directly to them, do not respond to stimuli from the skin, either tactile or chemical. Even when dropped into slowly-acting fixing solution, embryos of this stage do not give any evidence of muscular excitation.

With the ability to respond to tactile or chemical stimulation of the skin there appears a third series of cells. They bridge the gap

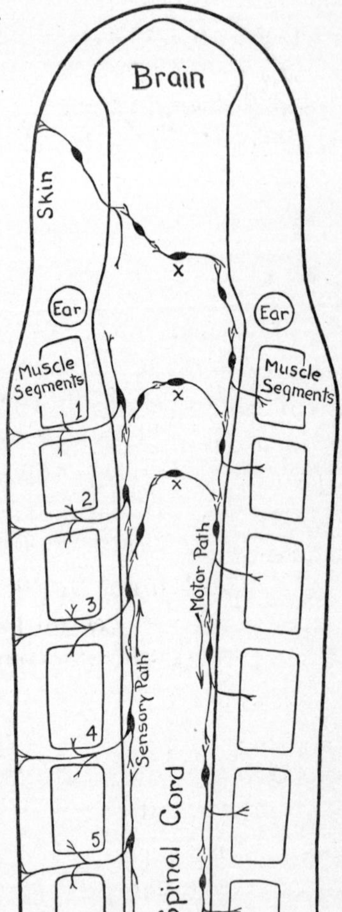

Fig. 10. A diagram of the mechanism which accounts for the coil reaction, cephalocaudal progression of movement and movement away from the side stimulated in *Amblystoma*. A stimulus in front of the ear would excite an afferent neurone, which, in turn, would excite a commissural cell in the floor plate. This cell turns caudad in the motor tract of the other side. A stimulus behind the ear, whether in the skin (exteroceptive) or in the muscle (proprioceptive) excites the afferent neurones which also excite the commissural cells of the floor plate, through which the impulse reaches the motor tract of the other side. Since the afferent system sends impulses to the motor system of the opposite side through floor plate cells and since the floor plate cells form a commissure only in the brain and anterior part of the spinal cord at this stage, impulses excited from any point in the body, whether headward or tailward, from the skin or from the muscles, must enter the neuromotor mechanism at its anterior end, and spread thence tailward. As a result of this, individual flexures progress cephalocaudad.

between the sensory system of one side and the motor system of the other (15). Their bodies lie in the floor plate of the medulla oblongata and upper part of the spinal cord (Figs. 9, 10). In the non-motile stage these cells are unipolar. The one pole of the cell extends either to the right or to the left into close relation with the motor tract on one side only. When they become bipolar they complete the path from the sensory field to the muscle; and this path leads to the muscles of the opposite side from the stimulus because the conductors from the sensory field pass across the motor path of the same side to establish synapses with the dendrites of the commissural cells in the floor plate (Fig. 9). It is found, accordingly, that the movements of embryos in the earlier period are typically away from the side stimulated. When this law appears to be violated and movement is toward the side stimulated there are probably multiple stimuli without precise localization.

Furthermore, since these commissural cells exist at this time only in the anterior end of the spinal cord and in the medulla oblongata, impulses from all parts of the body enter the motor mechanism in its anterior part, and from the anterior part proceed tailward through the descending or motor path (Fig. 10). The cephalocaudal progression of muscular contraction, which is the motive power of locomotion, has its origin in this definite and specific mechanism. But this mechanism is not adequate for locomotion in its typical form. It provides for reversal of flexures through the excitation of muscle-sensory nerve endings on the side of the initial contraction, and the conduction of that excitation to the muscles of the other side, that is to say, a contraction on the right can excite contraction on the left. But reversed flexures of this kind, although they may, when rapid, produce some locomotion, do not constitute the typical swimming. When this mechanism was first discovered it was regarded as the mechanism of typical swimming (28). Later observations, however, have brought to light another factor upon which aquatic locomotion depends.

THE SWIMMING MECHANISM

At the time swimming begins there is a growth of collaterals from fibres of the anterior part of the motor tract into relation with the dendrites of floor plate cells (Figs. 11, 25). These collaterals cause an excitation that is on its way to the muscles of one side to be carried through the commissural cells of the floor plate to the motor system of the other side. But in this passage to the muscles of the opposite side more synapses are involved than there are in the path to the muscles of the same side; so that the second flexure follows the first by a very brief interval. In the same manner as the impulse for the first flexure excited the second flexure, so the impulse for the second excites the third, and so on*.

As a factor in swimming, particularly in later stages, there may be an inherent rhythmic action of cells in the most anterior part of the motor system or in some more

* In response to the author's request, Dr G. P. McCouch offers the following comment upon the interpretation of the swimming mechanism as given above:

"During the coil stage, there are both exteroceptive and proprioceptive contralateral reflex arcs, but no ipsilateral arcs. These contralateral arcs consist of three series of neurones, all of which (from the trunk) are common to both exteroceptors and proprioceptors, which are supplied by branches of the same afferent neurones. It is probable, therefore, that exteroceptive and proprioceptive reflexes begin to remote part of the brain, for Detwiler has extirpated Mauthner's cells of *Amblystoma* embryos and found that the operated specimens do not acquire normal strength in swimming (22). But the animal can swim successfully with the nervous system transected at the level of the auditory vesicle. This means that the mechanism which is essential to the swimming movement is located virtu-

function at about the same stage of development. In that case, the coil stage may be characterized by two contralateral reflexes:

(1) A tonic postural proprioceptive reflex.

(2) An exteroceptive coil reflex.

"The slow onset and prolonged character of the coil reflex may be due not only to the early stage of development of the muscles and neurones, but also to the fact that the movement is mechanically opposed by the tonic contraction of the muscles on the side of the exteroceptive stimulus. The resulting increase in tension in the muscles of the convex side may set up proprioceptive stimuli which accentuate and prolong the coil reflex.

"The development of the 'S' reflex is associated with the growth of collaterals from the final effector axones which turn back to end about cells of the floor plate. The facts that in the mammal similar collaterals exist and that stimulation of the central end of a cut anterior root produces no contraction suggest that these recurrent collaterals are inhibitory in function. Their anatomic relations suggest that their reflex effects must be practically synchronous with those of the main branch of the axone from which they arise. At this stage, an exteroceptive stimulus gives an initial wave of flexion to the contralateral side of rapid onset and brief duration. The rapid onset may be due in part to reciprocal inhibition of the muscles on the side of the exteroceptive stimulus, which removes opposition to the initial flexion. The initial flexion stimulates a second reflex arising in receptors in the contracting muscle. This proprioceptive reflex excites the muscles of the convex side to which the effector neurones may be passing from inhibition to post-inhibitory exaltation (successive spinal induction). At the same time their recurrent collaterals inhibit the muscles of the flexed side, thus cutting short the initial flexion. In this way an exteroceptive reflex may be followed by a series of proprioceptive reflexes or swimming movements analogous to rhythmic stepping reflexes in the spinal mammal."

ally within the level of the muscle segments. In this part of the nervous system all that the animal can do, that is to say, its total behaviour pattern, is organized; and, as we have seen, it has become organized through a regular order of sequence of definite phases in the growth of the nervous system.

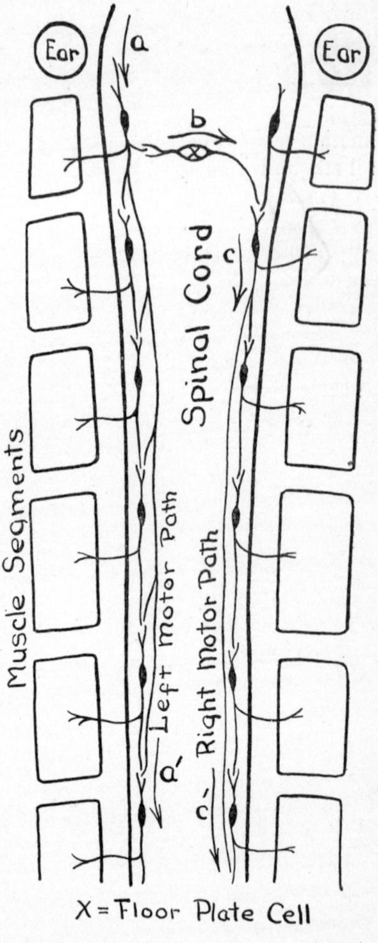

Fig. 11. A diagram of the neuromotor mechanism of swimming in *Amblystoma*. The sensory mechanism is omitted. The arrows indicate the direction of conduction. Arrow *a* represents the initial impulse which, as it passes tailward to *a'* and beyond, excites the muscle segments to a wave of contraction that progresses tailward. By the time the animal can swim, these neurones of the motor tract in the anterior region have developed collaterals which grow towards the median plane into synapse with commissural cells of the floor plate. This relation is illustrated at *b*, where the arrow indicates an impulse passing to the motor system of the other side. In the motor system it passes tailward according to the arrows *c* and *c'*. As explained in the text, this impulse, which has been excited by the collaterals, has been regarded as excitatory but it may be inhibitory (footnote, p. 14). Whether their function is excitatory or inhibitory, these collaterals are the essential structural addition to the mechanism as indicated in Fig. 1, and make it possible for the animal to use its muscular energy efficiently in locomotion.

TWO FURTHER GOALS OF BEHAVIOUR

At the time *Amblystoma* begins to swim it has no motile appendages (Fig. 5). There is still no mouth. There are external gills through which the blood circulates to serve the respiratory requirements, but these organs have as yet no power of movement. The fore limbs, which develop far in advance of the hind limbs, are at this time barely perceptible as nodules upon the surface. It is, therefore, several days after swimming begins before *Amblystoma punctatum* adds another feature to its behaviour pattern.

After locomotion by swimming is established, the development of behaviour in *Amblystoma* may be regarded as following two courses towards different goals: the one leading to the capture of prey and swallowing it; the other to terrestrial locomotion by walking. That these two courses are in a significant way physiologically distinct is shown by the fact that feeding and walking develop at very different relative rates in *Amblystoma punctatum* and *Amblystoma tigrinum* (axolotl). Movements of the limbs begin earlier in *Amblystoma punctatum* than in *Amblystoma tigrinum*, whereas movements of gills and jaws begin earlier in *Amblystoma tigrinum* than in *Amblystoma punctatum*. The course of development that leads to feeding involves the development of vision and movements of the jaws, branchial arches and related organs. The course that leads to terrestrial locomotion involves the limbs. Since the anatomical problem relating to the limbs is the simpler, it will be presented first.

MOVEMENT OF THE ARMS

The fore limbs develop much in advance of the hind limbs. The first movement of the fore limb is adduction and abduction. The limb, which projects tailward at an acute angle with the surface of the body, is drawn towards the body. When this movement of the limb is first performed it occurs only with trunk movement. When the trunk acts vigorously, as in swimming, the fore limbs are drawn close against the body (Figs. 12, 13). It has been observed critically that this is an active contraction and not a passive response of the limb to pressure against the medium in which the animal moves. This has been done by placing the animal in a small rectangular glass cage which prevents motion forward and consequent forced adduction of the limb as a result of pressure in forward movement. Under these conditions the animal cannot escape from the field of the microscope, and the movement of the limb is observed under sufficient magnification to make it clear in every detail.

LOCAL REFLEXES OF THE FORE LIMB

A day or two ordinarily elapses between the time when the arm begins to move with the action of the trunk before it acquires the ability to respond to a local stimulus without the perceptible action of the trunk (Fig. 14). Such independence of limb action appears to be acquired by a gradual reduction in the action of the trunk. At any rate, movement of the limb has been frequently observed to occur with slight movement of the head just before limb movement without perceptible trunk movement made its appearance. It is obvious, therefore, that the first limb movement is an integral part of the total reaction of the animal, and that it is only later that the

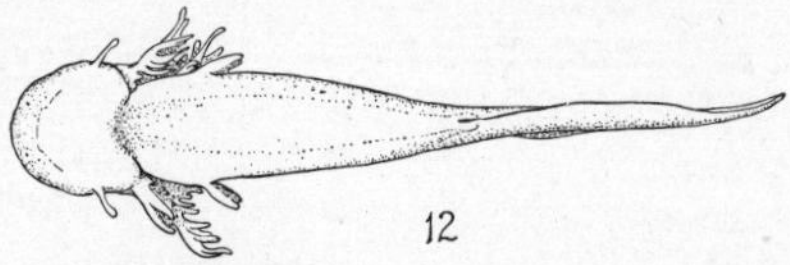

12

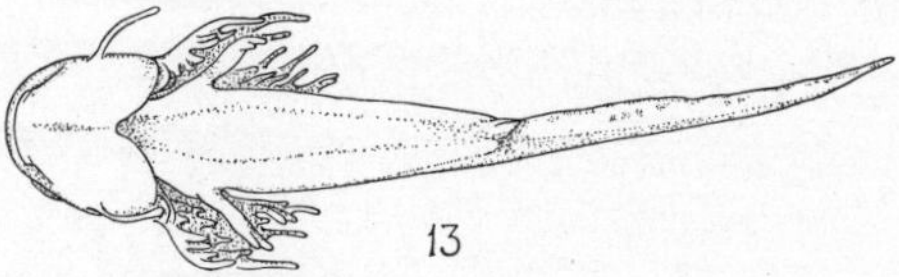

13

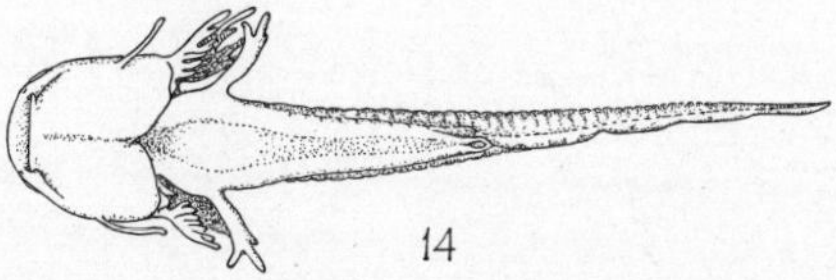

14

Fig. 12. Ventral view of *Amblystoma punctatum*, specimen 27 *B* 6, which moved its fore limbs slightly when the trunk moved but was negative to exhaustive test for both limb reflexes and gill reflexes. The gills moved with movement of the trunk.

Fig. 13. Ventral view of *Amblystoma punctatum*, specimen 27 *D* 51, which, in the supine position, responded with discrete movement of the fore limb; but in the earlier part of the test the limb movement was accompanied by slight movement of the trunk. Recorded as the earliest observed stage of discrete arm movement. The elbow flexed slightly when the limb as a whole moved. It is noteworthy that the reflexes could be evoked only when the animal was in the supine position. Magnification approximately the same as in Fig. 12.

Fig. 14. Ventral view of *Amblystoma punctatum*, specimen 26 *S* 9. The left arm of this animal gave a local reflex in response to a touch on the limb; but the right limb could not be stimulated to discrete local reflex. The position of the animal was upright. There was elbow flexion only with movement of the arm as a whole. Magnification approximately the same as for Figs. 12 and 13.

limb acquires an individuality of its own in behaviour. The local reflex of the arm is not a primary or elementary behaviour pattern of the limb. It is secondary, and derived from the total pattern by a process of individuation.

In the further development of the behaviour pattern of the arm the same principle is observed: the first elbow flexion occurs with action of the arm as a whole, and the forearm only later acquires the independence of a local reflex. So also is it with the movements of the hand and the digits.

That these early independent limb movements are identical with the definitive local reflexes with which physiology has so much to do is demonstrated by the behaviour of the limbs after the spinal cord has been transected above and below the spinal centres of limb movement. After an hour or more in physiological salt solution following such an operation, the limbs, forearm and fingers of *Amblystoma* of these stages under consideration will execute all possible movements in response to tactile stimulation. These reactions are exquisitely delicate and sensitive. It is obvious, therefore, that in *Amblystoma* the definitive local reflexes of the fore limb arise by a process of individuation out of the total behaviour pattern. The limb arises in absolute subjugation to the trunk. It can do nothing excepting as the trunk acts. From this subjugation it struggles, as it were, for freedom. The freedom which it ultimately attains, particularly under certain experimental conditions, has the appearance of being practically absolute, and the experimental reflex of this nature has come to be accepted as the elementary unit of behaviour.

Nevertheless, the local reflex of the fore limb of *Amblystoma* is not a simple or elementary thing in behaviour. It is a marvellously intricate, and derived product. This is demonstrated in performances which have been seen many times in the behaviour of *Amblystoma tigrinum* (axolotl). If a larva of this species, soon after the fore limbs move with trunk movement but before limb reflexes can be evoked with the animal in its normal upright position, is placed upon its back under conditions which twist the tail and trunk slightly upon the longitudinal axis, the fore limbs assume a definite posture. The limb on the side thrust deeper towards the bottom of the receptacle is then abducted and elevated while the other limb is adducted and relatively depressed. While in this posture, the extended limb will respond to a touch upon its surface by a further quick abduction and elevation.

THE RELATION OF POSTURE TO LOCAL REFLEX

In this performance there are two things of great importance: first, the earliest near approach to discrete limb movement appears as a postural reaction; and second, postural response sensitizes the exteroceptive reflex mechanism. In other words, the limb is able to respond very precisely to stimuli arising within the body (proprioceptive) as the result of a particular posture before it can respond to stimuli that arise exclusively from the outside world (exteroceptive); and this postural reaction of the limb enables it to respond to stimuli from the outside world before it can otherwise do so excepting as the trunk responds to the same stimulation. In this primitive vertebrate, therefore, the local reflex of the limb of the normal animal is primarily allied with posture. It is intricately involved from the very beginning with the behaviour pattern of the entire animal.

ANATOMICAL EXPLANATION OF DOMINANCE OF THE TRUNK OVER THE ARM

The secret of the sovereign power of the trunk over the limb as a rising dependency is found in the growth of the lines of communication. In our explanation of the nature of the movements of the trunk it was explained that nerve cells are arranged in a longitudinal series in the spinal cord in such a way that they conduct excitations from the head tailward, and that from these nerve cells side branches go to the muscle segments to excite them to contraction (Fig. 10). But these side branches, as motor nerves, do not stop growing when they reach the muscle segments. Having attached themselves to the muscle segment and established their control over it, they, by other branches, grow on beneath and beyond the muscle segment and invade the territory of the limb (Fig. 9). The first motor nerve fibres to reach the muscles of the limb are, therefore, branches of the same nerve fibres that stimulate the trunk muscles to action (18). Furthermore, these nerves reach into the territory of the limb-muscles long before muscle tissue is formed. As a result of this precocious invasion of limb-forming tissue by branches of nerve cells that are already integrating the trunk, the earliest movements of the limbs are of necessity totally integrated with trunk action. The central government, so to speak, establishes its sovereignty over the rising community before that community has acquired a central organization of its own, and subsequently grants to its subject more or less autonomy as time goes on.

HIND LIMB MOVEMENT

The hind limb acquires motility ordinarily ten or twelve days later than the fore limb (*Amblystoma punctatum*). Its earliest movements are, also, performed only as the muscles of the trunk contract (Figs. 15, 16). A day or so later it acquires the ability to execute reflexes without perceptible participation of the trunk in the response (Fig. 17). Also, when flexion of the knee first occurs it is a part of the movement of the leg as a whole. Only later does it occur in response to local stimulation without perceptible movement of the thigh.

The development of movements of the hind limb, therefore, follows the same order as has been described for the fore limb. Furthermore there is the same evidence here as in the case of the fore limb that these reflexes are identical with the definitive reflexes that are so well known to physiology, for they are all performed in response to tactile stimulation upon the leg or adjacent parts of the body after the spinal cord has been severed several segments in front of the pelvic girdle.

THE DEVELOPMENT OF TERRESTRIAL LOCOMOTION

The origin of aquatic locomotion has now been explained by the development of the primary motor mechanism in the nervous system. The complete subjection of the limbs to the trunk in their early activities has been explained by the fact that it is the swimming mechanism that makes the first nervous connection with the limb. This totally integrated relation of trunk and limbs proves to be the first principle of terrestrial locomotion.

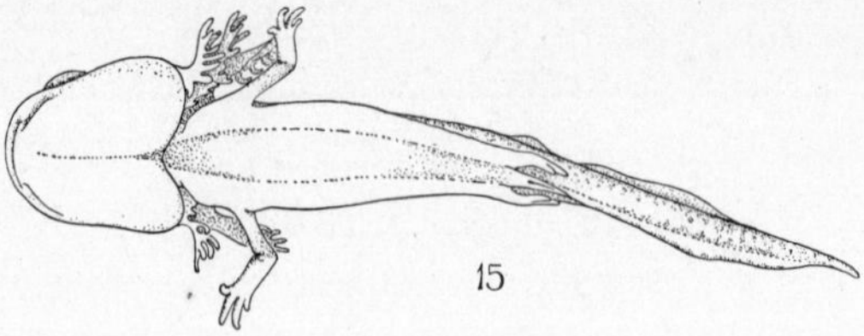

15

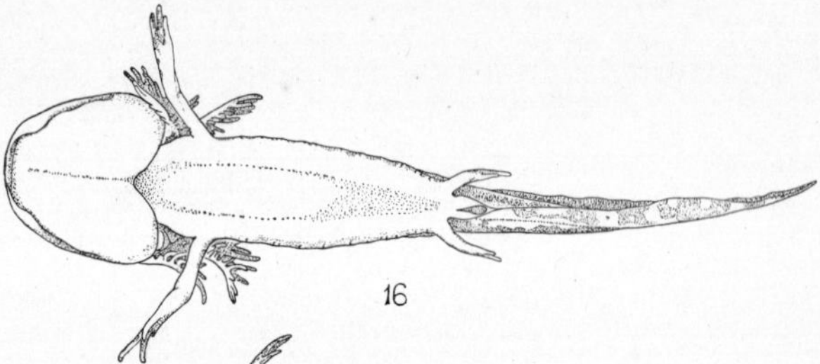

16

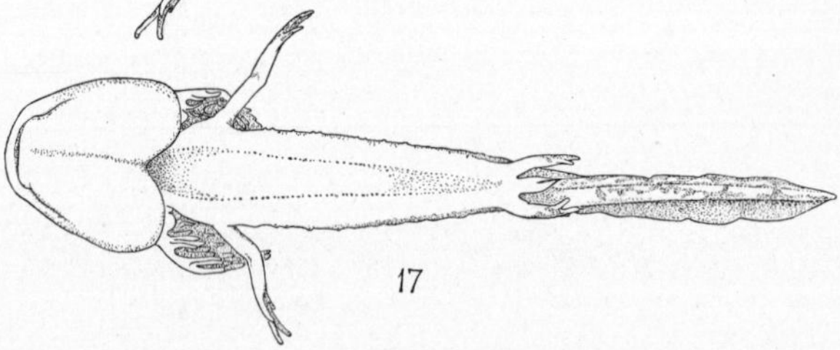

17

Fig. 15. Ventral view of *Amblystoma punctatum*, specimen 27 *F* 11, which abducted and adducted the hind limbs in conjunction with trunk movement but which could not be stimulated to give the local reflexes of the hind limbs. The animal lay perfectly quiet while the hind limbs and adjacent regions of the trunk and tail were stroked extensively with a hair, till finally the entire trunk moved.

Fig. 16. Ventral view of *Amblystoma punctatum*, specimen 26 *W* 13, of the latter part of the period in which the hind limbs move with trunk movement but do not move discretely. Magnification approximately the same as in Fig. 15.

Fig. 17. Ventral view of *Amblystoma punctatum*, specimen 26 *W* 15, which had been under close observation and was killed when it first responded to a stimulus on the hind limb with a local reflex. The action was adduction from a widely abducted position. It may have been due to an inhibition of abductor muscles. Magnification approximately the same as in Figs. 15 and 16.

As already stated, the fore limbs acquire motility far in advance of the hind limbs. They have in fact acquired all of their reflex activities before the hind limbs begin to move even under complete dominance of the trunk.

The first locomotor activity of the fore limbs is effected by the simultaneous action of both limbs. They are both abducted far forward and then adducted backward. This motion serves to move the body forward and to raise it slightly from the substratum. The head is usually elevated more or less at the same time. When this movement is well developed the head and shoulders are frequently raised high upon the fore limbs at full length and from this position the animal falls forward. Sometimes this elevated position of the body is maintained for a brief period and then the head is turned laterally as if to explore the field visually (Fig. 18). From such an attitude as this the animal has been seen to strike into its first walking gait. This is done by flexure of the anterior part of the trunk laterally and abduction of the arm on the convex side with adduction of the arm on the concave side (Fig. 21).

The first alternate movements of the fore limbs in walking are therefore integral parts of trunk movement. The primary co-ordination in the performance is not directly between the two limbs. Each limb, on the contrary, is directly integrated with its own side of the trunk. Furthermore, movement of the trunk in walking is nothing more nor less than the swimming movement with greatly reduced speed. It is a slow sinuous flexure progressing from the head tailward.

Early in the period of walking only with the fore limbs, the hind limbs are drawn forward (abducted) simultaneously and held in the extended position while the fore limbs walk. This occurs before there is flexion of the knee. Typical co-ordination of all four limbs in the initial movement of walking, extension of the fore limb of one side with the hind limb of the other, occurs at about the time local reflexes appear in the hind limbs; but it has been seen in specimens in which local reflexes could not be elicited (Fig. 19). The co-ordination of the hind limbs in walking is, therefore, not dependent upon well developed local reflexes (Fig. 20).

In its earliest performance the complete co-ordination of the walking gait may be sustained only for the initial movement, after which the hind limbs drop out of the gait while the fore limbs continue the process. In this initial co-ordination of the hind limbs for walking, the trunk is bent laterally and the hind limb on the concave side is abducted while that on the convex side is adducted (Fig. 22). The hind limbs are, therefore, integrated with the trunk in the beginning of walking upon exactly the same principle as are the fore limbs; except that while the fore limb is adducted on the concave side of the trunk the hind limb on that side is abducted. But the hind limbs do not act simultaneously with the fore limbs. They lag behind the fore limbs with an interval corresponding to the time involved in the progression of the trunk flexure from the brachial to the pelvic region.

The swimming movement is, therefore, the initial and dominant factor in terrestrial locomotion, and the earliest action of the limbs in this function is under absolute dominance of the trunk. Walking does not come about through the co-ordination of local reflexes of the limbs. Each limb in the early performance is integrated with

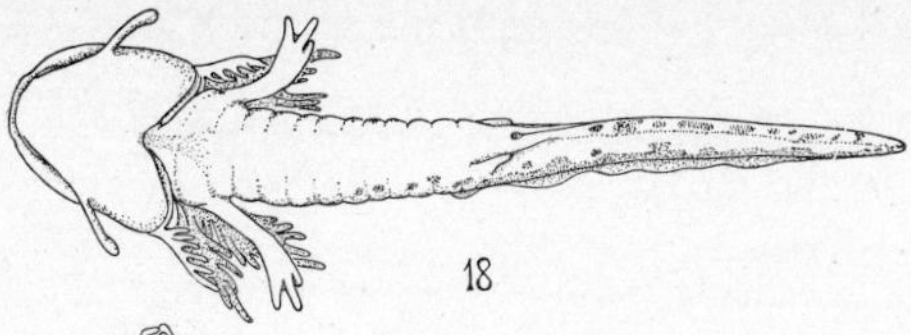

18

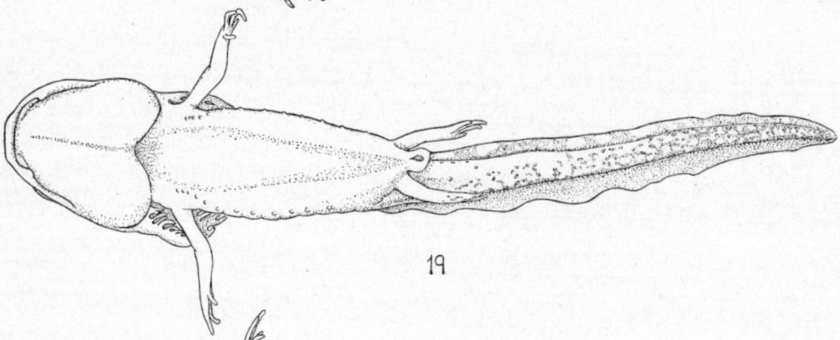

19

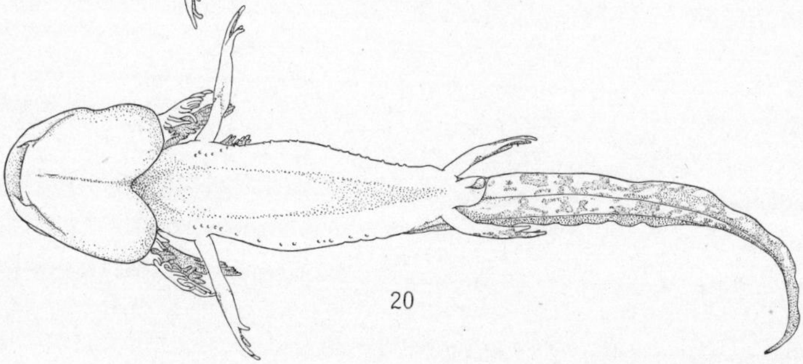

20

Fig. 18. Ventral view of *Amblystoma punctatum*, specimen 27 *F* 19, which was selected as a type of the stage of earliest walking movements of the fore limbs. The animal slowly turned its head to one side with both feet on the substratum and head and shoulders elevated. The head and fore part of trunk then turned to the right and simultaneously the left foot stepped forward in typical walking movement. The alternateness of limb movement was in this case distinctly integrated with the movement of the trunk, or an integral part of a total reaction; the antigravity component of the performance may have been a local reflex.

Fig. 19. Ventral view of *Amblystoma punctatum*, specimen 26 *I* 3, in which local reflexes of the hind limbs could not be excited, but which walked strongly with typical co-ordination of all four limbs. There was no flexion of the knee in walking. It did not even bend readily under pressure. Magnification approximately that of Fig. 18.

Fig. 20. Ventral view of an *Amblystoma punctatum*, specimen 26 *M* 2, which was selected as the earliest stage of discrete flexion of the knee without movement of the leg as a whole. This reaction occurred in the right leg but could not be evoked in the left. Magnification approximately the same as in Figs. 18 and 19.

the trunk. As one flexure after another passes from the head tailward, the limbs respond to the dominance of the trunk over them just as they did in their earliest movements.

Now, it has been observed that the fore limb and hind limb acquire apparent independence of action in local

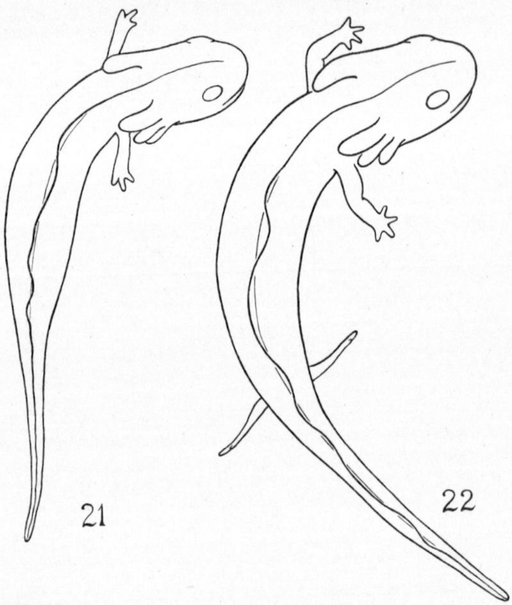

Fig. 21. Diagram illustrating the initial act of walking with the fore limbs by specimen 27 *F* 19, illustrated in Fig. 18.

Fig. 22. Diagram of initial act of walking with all four limbs by *Amblystoma* of the stage represented by Fig. 19, when the hind limbs have no power of reflex action in response to local stimulation and no knee flexion.

reflexes by a gradual reduction in the magnitude of the dominating trunk movement. So is it in the development of the apparent independence of the limbs in walking; the sinuous movement of the trunk becomes more and more reduced in magnitude relative to the dimensions of the animal and the limbs acquire more and more an apparent independence of action. Finally, walking appears to be the result of co-ordination of all four limbs as essentially independent reflex mechanisms. But, just as in the case of the local limb-reflex and its relation to postures of the trunk, so, in walking, there is the individuation of parts within the total pattern and not integration of essentially or primarily independent units.

THE MECHANISM OF CO-ORDINATION IN WALKING

Since, as just explained, the swimming movement is the primary factor in walking, and since the limbs are at first dominated by the trunk in walking as they are in their earliest movements, the mechanism that explains the swimming movement and the early dominance of the trunk over the limbs must be the essential mechanism of terrestrial locomotion. Although great development has taken place in the brain and spinal cord of *Amblystoma* between the time when it begins to swim and the time when it begins to walk (Fig. 52), many motor root fibres still emerge from the same longitudinal tract from which they arise in the early swimming stage. It is impossible at present to assert that there are fibres from no other source, possibly of more local origin, in the motor nerve roots of the walking animal; but it is certain that there is a large component in these roots that corresponds exactly with the fibres that excite the early swimming movement and establish the dominance of the trunk over the limbs. The enlargement and extension of this mechanism explains the initial co-ordination of walking. The mechanism by which greater independence of limb action is brought about awaits further investigation. It will receive general consideration in the third lecture.

MOVEMENTS OF THE GILLS

The first new component of the behaviour pattern to make its appearance after *Amblystoma* begins to swim is movement of the external gills. But when these movements are first performed, and for a considerable time in the more slowly developing species, they occur only with movements of the head or trunk; the gills are depressed when contractions occur in the muscles of the trunk (Fig. 23). Only later can movement of the gills as a local reflex be excited (Fig. 24). Such reflexes occur in response to light touch on the base of the gills, the top of the head, and, with particular facility, to touch on the fore limb or its immediate vicinity. It is an important feature of this local reflex that it is evoked by a very weak stimulus, a very light touch, while a strong stimulus on the same region or a light touch on other regions excites a total response of trunk movement and gill movement. As in the case of the limbs, the gill reflex, by reduction in the trunk component of the movement, attains discreteness gradually.

THE ANATOMICAL EXPLANATION OF DOMINANCE OF THE TRUNK OVER THE GILLS

The close alliance of the gill-movement with trunk movement has particular interest because the muscles which move the gills receive their nerve supply from a quite different centre from that which supplies nerves to the trunk muscles. Their nerves come from the centres that control the muscles of the jaws and branchial arches, in the functions of mastication and swallowing. The anatomical explanation for this primary alliance of the gill must therefore be quite different from that for the dominance of the trunk over the limb. The first nerve

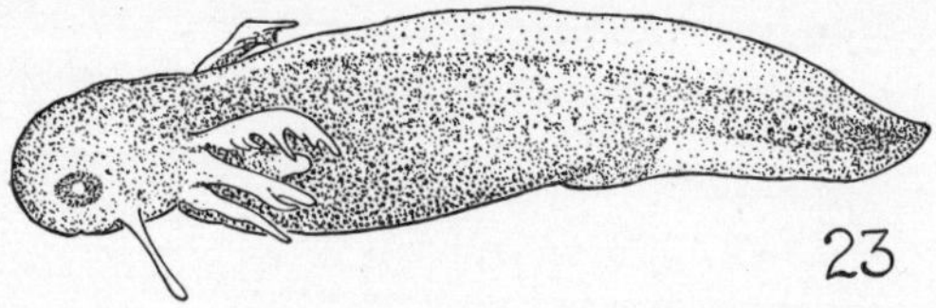

23

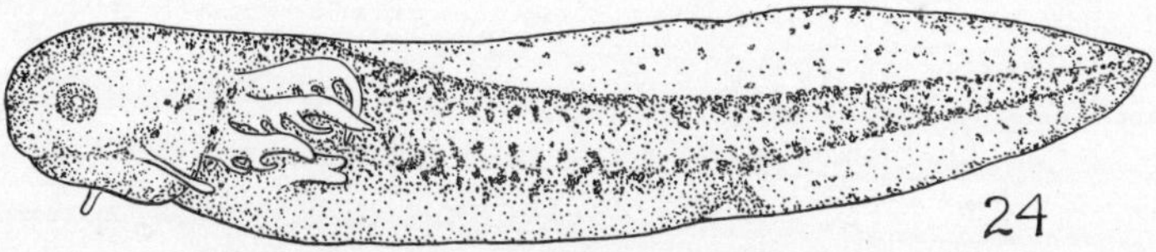

24

Fig. 23. A dorsolateral view of an *Amblystoma punctatum*, specimen 28 *B* 2, which depressed the gills with movement of the trunk but could not be stimulated to move the gills discretely in response to a local stimulus. The threshold of tactile stimulation upon and near the gills and upon the fore limb was very high. The limbs could be forcibly bent extensively without exciting movement of any kind. The specimen is foreshortened in the figure by lateral curvature.

Fig. 24. A lateral view of an *Amblystoma punctatum*, specimen 27 *D* 6, which gives prompt gill reflexes in response to touch: 23 hours previously it moved the gills only with trunk movement. In the stage figured the limbs did not move with trunk movement. Two days later specimens of the same lot began to snap at *Daphnia*. Magnification approximately the same as in Fig. 23.

fibres to the limbs, as just explained, are branches of the nerve fibres to the trunk musculature, whereas the nerve fibres to the muscles of the gills arise from cells that have no contacts with the muscles of the trunk.

In a transverse section of the medulla oblongata of *Amblystoma* of the early swimming stage, the motor cells of the group which are to innervate the muscle of the gills are situated in the most dorsal part of the motor division of the nervous system (Fig. 25)(17). Large dendrites from these cells invade all sources of afferent impulses through the sensory system, the ascending trigeminal tract and the lateral line tract. They reach also into the zone of synapse between the commissural cells of the floor plate and the motor neurones that innervate the trunk muscles. The muscles of the external gills are therefore subjected to excitation not only by those afferent impulses which are on the way to the motor mechanism of the trunk, but also by those efferent impulses of the motor path which are on the way directly to the muscles of the trunk. In general, therefore, by reason of these relations of the motor neurones that innervate the muscles of the gills, these muscles, after they have acquired sensitiveness to nervous excitation, would be expected to receive excitation from impulses that are on the way to the muscles of the trunk.

Since the gills act with the trunk before they act independently in response to afferent impulses it must be that the motor neurones that control them establish functional synapses with the motor neurones that excite trunk movement before they establish such relations with the afferent neurones. It has not been demonstrated in detail that this occurs, but it has been demonstrated beyond question that throughout the part of the brain concerned dendrites of all motor cells grow ventrad before they grow dorsad. It is, therefore, reasonable to conclude that the ventrally directed dendrites are the first to become functional conductors. Accordingly, since the sensory field of the brain is dorsal of the motor cells that control the gills, and the motor tract that controls the trunk is ventral, there is ample anatomical reason for the earliest movements of the gills to be excited by the motor system that excites trunk movement.

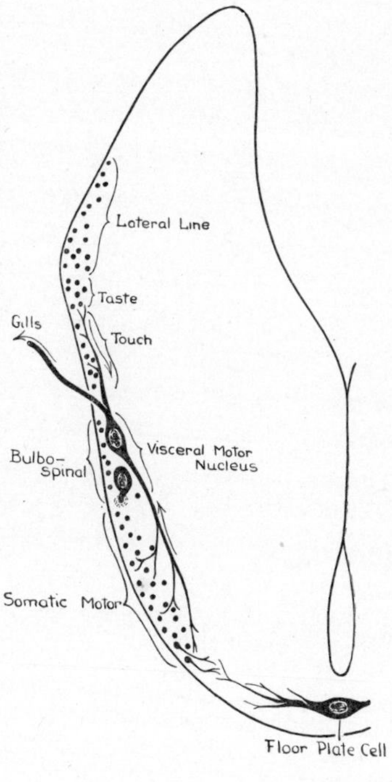

Fig. 25. A diagram to illustrate the mechanism that causes the earliest movements of the gills to be integrated with the trunk instead of appearing as local reflexes. Cells of the visceral motor nucleus innervate the muscles of the gills by their axones and by their dendrites reach ventrad into the somatic motor path that controls the muscles of the trunk. These dendrites arise earlier, and therefore are regarded as becoming functional earlier, than the dendrites which reach dorsad into the sensory tracts which represent touch and taste. The bulbo-spinal path arises from neurones of the visceral motor nucleus and serves to integrate the several nuclei of this system (Fig. 28) in the same manner that the somatic motor paths (fasciculus longitudinalis medialis, Fig. 29) integrate the somatic motor nuclei. The collaterals that are pictured as going from the fibres of the somatic motor path into relation with floor plate cells represent the collaterals described in the text as essential to the swimming movement. The degree of differentiation shown in the figure represents the early swimming stage at the level of the glossopharyngeal nerve.

THE FEEDING REACTION

The capturing of food is a complicated performance, for it involves action of the trunk, jaws and branchial arches. Its early development indicates that the trunk component, consisting of a short quick jump forward, becomes functional first. This characteristic movement can be evoked by a very light touch on the limb or adjacent parts of the body before there is any evidence of visual responses. Somewhat later it occurs when a hair or bristle is moved in the visual field at a distance of two or three millimetres from the eyes (Fig. 26). The animal then jumps forward at the moving object, but without making any perceptible jaw movements (Fig. 27). This is an exceedingly quick action and the animal

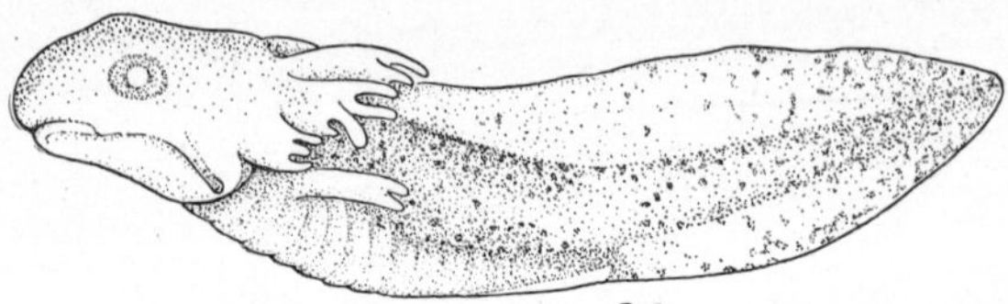

26

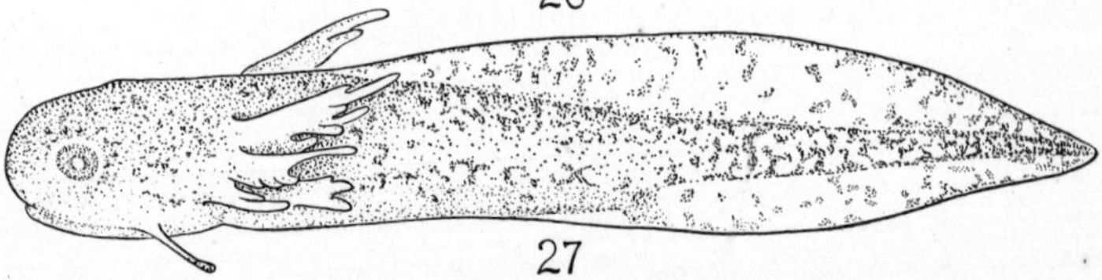

27

Fig. 26. Lateral view of an *Amblystoma punctatum*, specimen 27 *F* 25, in which conjugate deflection of the eyes was observed. Reflexes in the fore limbs could not be elicited. This is approximately at the time feeding begins. (Compare Fig. 27.) The specimen is foreshortened by lateral curvature.

Fig. 27. Lateral view of an *Amblystoma punctatum*, specimen 26 *O* 2, in the early feeding stage. The limbs, balancers and gills appear to be slightly more advanced in development than those of specimen in Fig. 26. The latter was selected for early conjugate deflection of the eyes.

comes to rest immediately. Later the performance involves actual snapping at the object that is moved in the field of vision. This is certainly a visual response, for it occurs when there is no possibility of tactile excitation.

ANATOMICAL EXPLANATION OF THE FEEDING REACTION

The explanation for the precocious development of the trunk component of this response must be in the early development of the fasciculus longitudinalis medialis from the midbrain to the motor centres of the trunk (Fig. 29). This tract is well developed, together with the posterior commissure, in the early swimming stage, when the optic nerves are just entering the brain in their growth from the retina. It must respond to the earliest feeble visual impulses, and conduct resultant impulses from the optic fibres directly to the motor path of the trunk. It must, in response

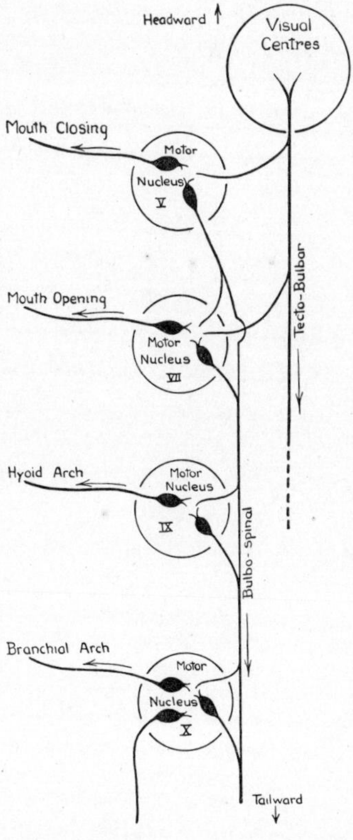

Fig. 28. A diagram representing the visceral motor centres that are involved in feeding as they are arranged along the longitudinal axis of the nervous system. This interpretation is based on the structures observed in the early swimming stage or possibly in specimens a few hours past that stage. Mouth opening and closing are antagonistic, and would require inhibition of the one with excitation of the other. The tecto-bulbar path, which can be recognized in the early swimming stage, is probably primarily concerned with the feeding mechanism, and the integration of the mechanism as such must be effected through the bulbo-spinal path. An impulse passing tailward through this system would give rise to movements of parts in the order observed in the feeding reaction. This figure should be studied in relation to Fig. 29.

to visual stimulation, excite the quick, short trunk movement which initiates the feeding reaction.

The jaws and branchial arches become allied with this initial trunk movement of the feeding reaction through a mechanism that is essentially that which allies the external gills with movements of the trunk, that is to say, dendrites of the motor neurones that control the jaws and branchial arches engage in synapse with the motor neurones that excite the action of the trunk (Fig. 29). As an excitation passes tailward through the fasciculus longitudinalis medialis, the motor nuclei of the trigeminal, facial, glossopharyngeal and vagal nerves are excited to action in regular order so that the jaws, hyoid arch and branchial arches act in appropriate sequence and simultaneously with the action of the trunk. Swallowing is initiated by action of the branchial arches; but young *Amblystoma* snap at their prey and frequently kill it before swallowing is possible. The function of ingesting food is therefore dependent finally upon the development of functional relation between the muscles of the gullet and the vagus nerve. The ontogenetic development of the various components is therefore in the order of trunk, jaws, branchial arches, oesophagus (Fig. 29). This is also the order of physiological actions.

SUMMARY

By way of summary several conclusions may be drawn from the results which we have here presented of our studies upon *Amblystoma*.

1. The behaviour pattern develops in a regular order of sequence of movements which is consistent with the order of development of the nervous system and its parts.

2. In a relatively precise manner physiological processes follow the order of their embryological development in the functions of aquatic and terrestrial locomotion and feeding.

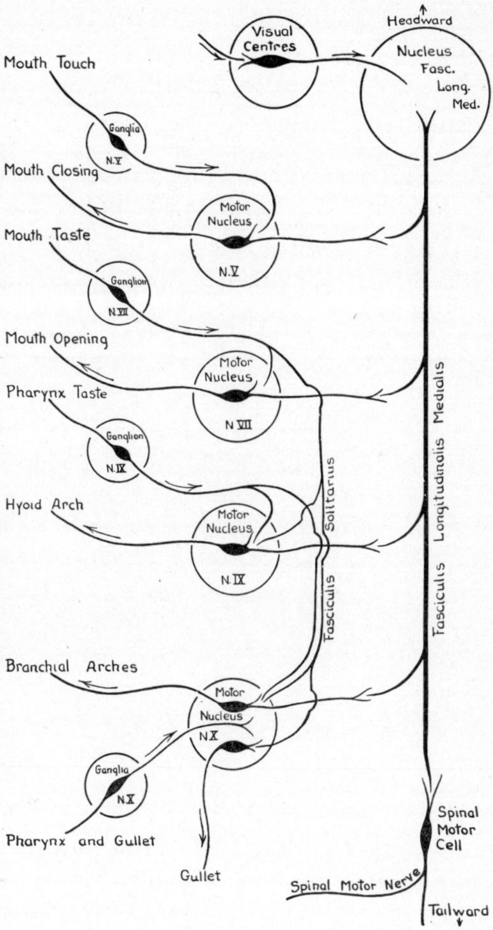

Fig. 29. A diagram supplementing Fig. 28 for illustration of the mechanism of feeding. The feeding reaction, stimulated from the visual field, is initiated by trunk movement, which is obviously brought about through the action of the fasciculus longitudinalis medialis upon the spinal motor system. As illustrated in Fig. 25, this tract is invaded by dendrites of the motor neurones of the nerves V, VII, IX and X. The original impulse passing from the optic centres to the spinal motor nerves probably integrates this mechanism as a whole under the action of distant vision; whereas close vision, as the prey is approached, may act through the tecto-bulbar system (Fig. 28) to inhibit jaw closing through collaterals and excite mouth opening through axone terminals. Touch in the mouth probably excites jaw closing, and touch and taste in the mouth and pharynx probably excite action of the hyoid and branchial arches and finally the gullet. With the cessation of the visual stimulus the somatic motor system becomes relatively quiescent and the bulbospinal system probably effects the appropriate inhibitions and excitations for the act of swallowing just as the somatic motor system accomplishes this for swimming.

The act of feeding obviously begins as a total reaction with subsequent physiological individuation of the swallowing mechanism largely through strong excitation from the visceral receptor field.

The anatomical relations illustrated in this diagram as pertaining to the central nervous system are based on conditions found in the early swimming stage. At this time the peripheral parts of the nervous mechanism of swallowing are established in varying degree, but the muscles involved have not yet differentiated out of the mesenchyme. The remarks on excitation as opposed to inhibition at particular points in the mechanism are inferential.

3. Behaviour develops from the beginning through the progressive expansion of a perfectly integrated total pattern and the individuation within it of partial patterns which acquire various degrees of discreteness.

The last conclusion will receive consideration in the third lecture particularly. The next lecture will deal with the origin of the nervous mechanism which, as explained in the present lecture, dominates behaviour in its early development.

Lecture Two

DYNAMIC ANTECEDENTS OF NEURAL MECHANISMS

OUR first lecture dealt exclusively with the behaviour pattern of a particular animal and the explanation of that pattern upon the basis of functional nervous mechanisms. It is our purpose now to consider the origin of those nervous mechanisms from structures and functions that antedate nervous conduction in the development of the individual.

The embryo is perfectly integrated before it has a nervous system. Not only are the processes of growth within its tissues co-ordinated in such a way that development of all parts progresses normally, but there is, in *Amblystoma* and many other Amphibia, a co-ordination of ectodermal cilia, the rhythmic motion of which keeps the embryo in constant movement within the egg membrane. Between this earlier non-nervous and the later nervous condition of the embryo there appears superficially to be a hiatus in development, a point where one condition ends abruptly and something entirely different takes its place. How is this transition made from the non-nervous mode of integration to the nervous? Obviously, in attempting to answer this question, the first item to consider is the nature of the process of integration that prevails before the appearance of nervous conduction.

FORCES OF INTEGRATION IN LOWER ORGANISMS

The basic factor of organization which here requires consideration is polarity. The animal egg expresses polarity in the differentiation of poles known as animal and vegetative, or apical and basal. From this condition it is transformed into an organism with oral and aboral or anterior and posterior poles. Among the tissues as development proceeds gland cells become polarized so as to discharge their secretions in a fixed direction, muscle cells become polarized so as to pull on definite points, and nerve cells become polarized so as to conduct their impulses from dendrites through axones to precise destinations. It is this principle of polarity of the organism and its tissues that enables the animal to direct its forces in such a way as to put itself in appropriate relation to its environment.

The polarity of an animal is not simply a matter of structural or ordinary physiological differences between parts. It is the expression of forces in action, that is to say, it is dynamic. This was originally demonstrated in the discovery of the variable susceptibility of protoplasm to the lethal action of various substances*. The oral end of the polyp of a hydroid, for example, disintegrates more rapidly than the aboral end under the action of potassium cyanide or other poisons. This graded intensity of action from one point to another in the tissues

* Upon the subject of susceptibility gradients there is an extensive literature which cannot receive special attention in this lecture. It has been recently summarized and discussed by Child, who gives, also, a comprehensive bibliography upon the subject (10).

is designated a susceptibility gradient. This gradient of susceptibility to chemical reagents as it appears in many organisms is now known to be coincident with gradients of susceptibility to temperature and radiations of various kinds, and gradients of carbon dioxide production, oxygen consumption, oxidation-reduction phenomena and electrical potential as well as of secretion, contraction and conduction. In their various aspects these phenomena are in general called metabolic gradients (7, 8, 9, 10).

But this dynamic polarity of organisms or tissues is not a fixed thing and inherent in their structure. It is a relative thing and can be determined or modified by extrinsic factors. It is found, for example, that in the regeneration of a piece cut out of the middle of a hydroid polyp, the end which lay nearest the oral end in the original animal will become the oral end of the new animal; and the aboral region of the piece will become the aboral pole of the new organism. If, however, such a regenerating piece is subjected appropriately to an electrical current or other adequate stimulation, this polarity can be reversed so that the end of the piece that would normally become the oral end can be made the aboral, and the polarization thus established has all the characteristics of the polarity that develops without interference (38).

As already indicated, this polarity of gradients which can be activated and determined electrically, expresses itself also in a differential electrical potential. This potential is such that the oral end of the polyp is electronegative to the aboral end. Polarity of the organism, therefore, expresses itself, at least in part, in purely physical modes, which are themselves activating agencies of physiological and growth processes within the organism. It is this conception of an organization of intrinsic forces that are accelerated and retarded in their interaction upon one another that we wish to apply to the interpretation of the transition from the non-nervous to the nervous mode of integration in the growth of *Amblystoma*.

PRENEURAL GRADIENTS IN *AMBLYSTOMA*

The application of the susceptibility methods of experimentation to the unsegmented eggs of vertebrates shows that these eggs are polarized similarly to the hydroid polyp in that there is a metabolic gradient with its "high" end at the apical or animal pole and its "low" region about the basal or vegetative pole (10). At about the time that the embryo of the frog or of *Amblystoma* begins to elongate from the spherical condition of the egg the metabolic gradient shifts with the axis of the animal and becomes longitudinal. It appears in the ectoderm and nervous system with its region of higher rate in the anterior end (Fig. 30). This appears to hold for all vertebrate embryos that have been investigated in this manner. Observers generally find also, a little later in development, a centre of relatively high rate in the caudal end of the embryo, and assign this centre to the nervous system or to the nervous system and the mesoderm. According to my own observations, when an early embryo of *Amblystoma* is placed in an appropriate solution of potassium cyanide, the nervous system first begins to disintegrate in the head region and the disintegration progresses tailward, whereas the mesoderm begins to disintegrate in the caudal region and its disintegration progresses headward. Apparently the nervous system is also involved to some degree in the degeneration at the caudal end.

This progression of disintegration from the caudal part of the mesoderm headward is best observed by cutting the embryo open from the ventral side and re-

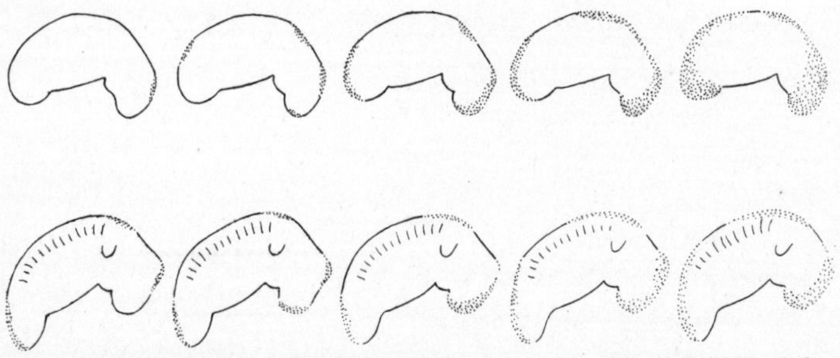

Fig. 30. Two embryos of *Amblystoma microstomum* showing (from left to right), by stippled areas, the progressive disintegration of the tissues under the action of potassium cyanide, as indicative of metabolic gradients. In the upper specimen disintegration begins at about the level of the diencephalon, and very soon thereafter in the most rostral part of the forebrain, in the hindbrain and near the caudal end of the spinal cord. The anterior three regions soon fuse, while the caudal region becomes more extensive. Finally the anterior and posterior disintegration areas meet in about the middorsal region.

In the lower specimen of the figure disintegration seems to begin in the caudal region as early as in the anterior region; but it soon becomes more active in the anterior region.

moving the entoderm. The ventral face of the mesoderm may then be seen clearly and exposed directly to the action of the cyanide solution, without such obstructions as the skin offers when the solution acts from the outside of the embryo. Under these conditions the mesoderm has been seen to disintegrate, first, in the unsegmented region and then progressively headward, one myotome after another. The more rostral segments are found to be much more resistant to the reagent than are the unsegmented mesoderm and the more caudal segments.

It appears, therefore, that there are two longitudinal metabolic gradients operating in the embryo of *Amblystoma* at the time when the conducting mechanism of the nervous system is beginning to take form. The one of these gradients operates in the nervous system, the other in the mesoderm (Fig. 31). In view of the extensive knowledge now current concerning metabolic

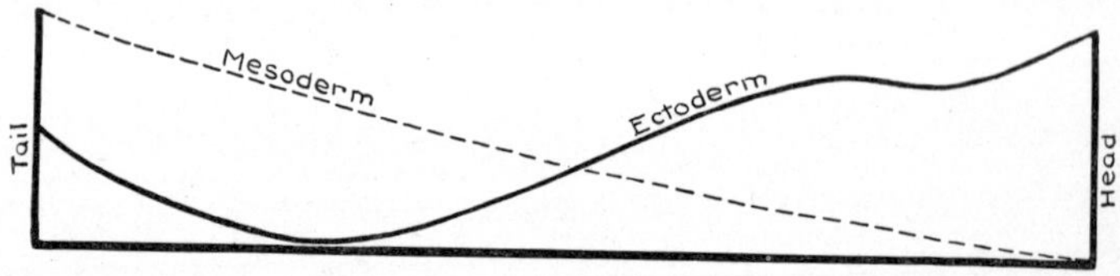

Fig. 31. Graphs to illustrate the ectodermal and mesodermal metabolic gradients as they occur in *Amblystoma* according to Fig. 30. In the ectoderm the metabolic rate appears in general to be highest in the forebrain. In the hindbrain and possibly the rostral part of the spinal cord there is another centre in which the rate seems to be higher than just in front of it. In the midtrunk region the rate appears to be lowest, while in the extreme caudal region in early stages the rate seems to be slightly higher than in the midtrunk region. The quantitative aspect of these graphs is purely hypothetical. The probability is that the graph representing the mesodermal gradient should fall rapidly from the tail end headward through the extent of the unsegmented mesoderm, and then run at a low level to the head region.

gradients in lower organisms, it is a reasonable hypothesis that these gradients in *Amblystoma* are expressions of the integrating forces of the organism as a whole. They are physiological or dynamic antecedents of neural functions. But in order to show how two such systems of forces may be operating simultaneously within the embryo and acting upon its tissues in such a way as to be factors in the origin of neural conducting mechanisms, one must picture in considerable detail the structural relations of the parts that are involved.

ANATOMICAL RELATIONS OF THE PRENEURAL GRADIENTS

While *Amblystoma* is still in the spherical form of the egg the central nervous system makes its first appearance as a thickening of the ectoderm, or primitive skin, over the dorsal surface of the embryo. This thickening rises along its lateral margins into ridges which are approximately parallel through the region which will become the trunk (Fig. 32). In a transverse section of the embryo at this stage, the central nervous system is seen as the medullary plate with its thickened margins sharply differentiated from the thin skin over the rest of the body (Fig. 33). The ridges at the margins of the medullary plate are the neural crest folds. The inner face of the plate is fairly smooth and even throughout excepting for a broad depression affecting the middle portion and making the plate thinner in this region. A membrane is now forming clearly on the inner face in the lateral portion of the plate, but it is indistinct in the middle portion. This becomes the external limiting membrane of later stages. The middle portion of the medullary plate at this time rests upon the entoderm, and the latter is folding up against the plate to form the

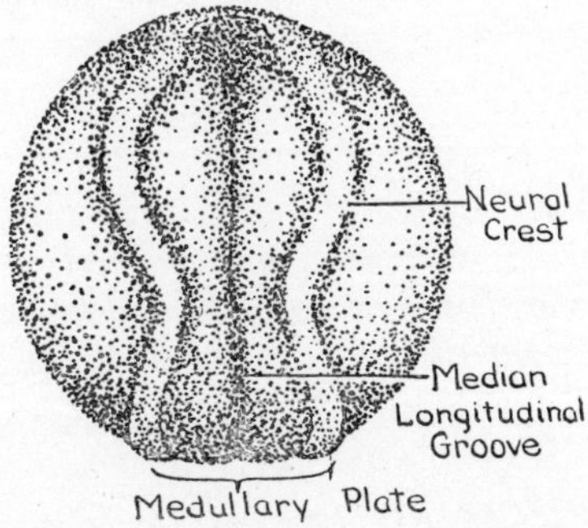

Fig. 32. A dorsal view of an embryo of *Amblystoma punctatum* in the medullary plate stage. Figs. 33 to 36 represent sections through the medullary plate of about this stage.

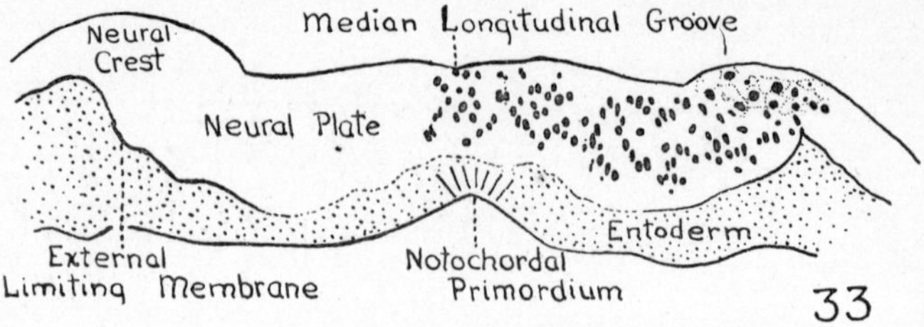

33

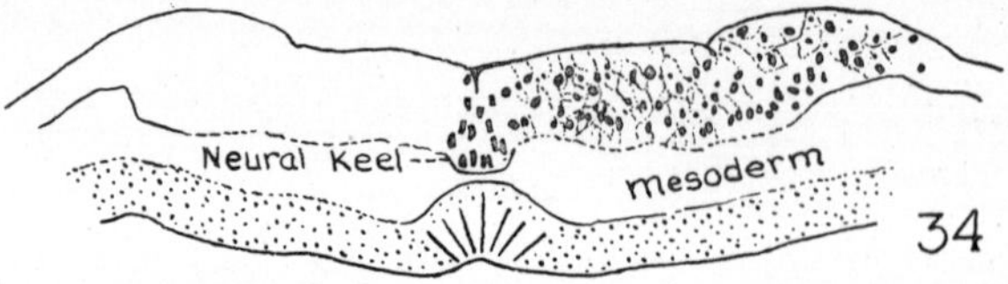

34

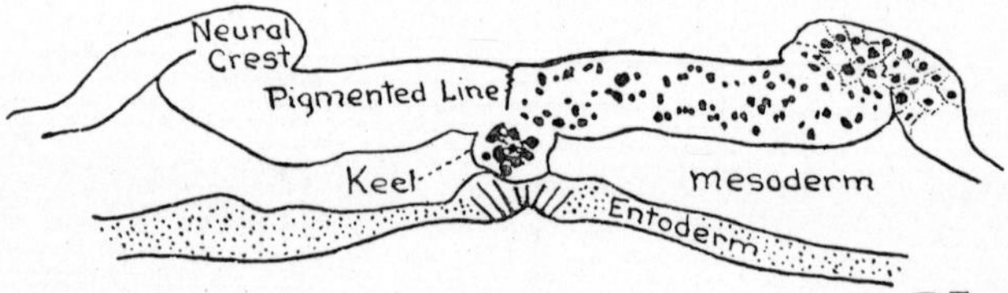

35

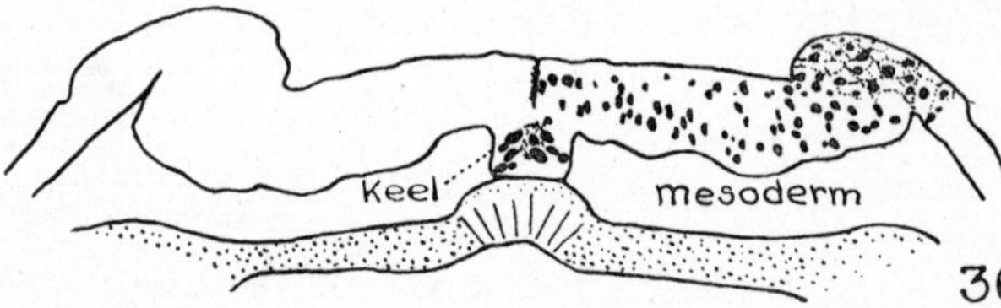

36

Figs. 33 to 36. Sections through the medullary plate and subjacent structures of *Amblystoma*. They are respectively copies of Baker's Figs. 4, 7, 8 and 9, with modifications only in the labelling (2). In Fig. 33 the medullary plate rests upon the entoderm. In Figs. 34, 35 and 36 the mesoderm has grown in between these two structures as far as the neural keel and the primordium of the notochord. The migration of cells towards and into the neural keel in Figs. 33 and 34, and their radial orientation in Figs. 35 and 36 are noteworthy. The progressive development of pigment from the median longitudinal groove inward, and in the deeper part of the neural keel, is referred to in the text.

beginning of the notochord. Only the most lateral part of the plate faces the mesoderm.

In the course of a few hours the medullary plate undergoes important changes on its inner face. Immediately over the primordium of the notochord the middle part pushes inward to form a ridge, which rests directly upon the notochord. This ridge was discovered by Baker and called by him the "neural keel" (Fig. 34)(2). There is now a distinct external limiting membrane across the ventral face of the neural keel. Also the mesoderm has crowded in from the lateral regions and filled snugly the whole space between the entoderm and the medullary plate. It is at this time a continuous sheet of tissue without any evidence of segmentation into myotomes. Its most massive part lies against the neural keel.

While the keel is in this typical condition the margins of the plate fold upward to form the neural tube, and as the lips of the groove come together the cells of the neural crest break loose from the neuroepithelium and migrate into the space between the skin and the neural tube (Figs. 36–40). At this time the central nervous system touches the mesoderm only along the ventral surface of the neural tube and the sides of the neural keel. The entire dorsal and lateral wall faces the skin directly. The skin even grows into the crevice between the tube and the mesoderm, forming a thickening which constitutes, as it were, a barrier between the mesoderm and the dorsal or sensory portions of the neural tube.

These facts of anatomy show that the main mass of the mesoderm in which the mesodermal metabolic gradient operates is intimately related to the medioventral part of the nervous system; and that the latero-

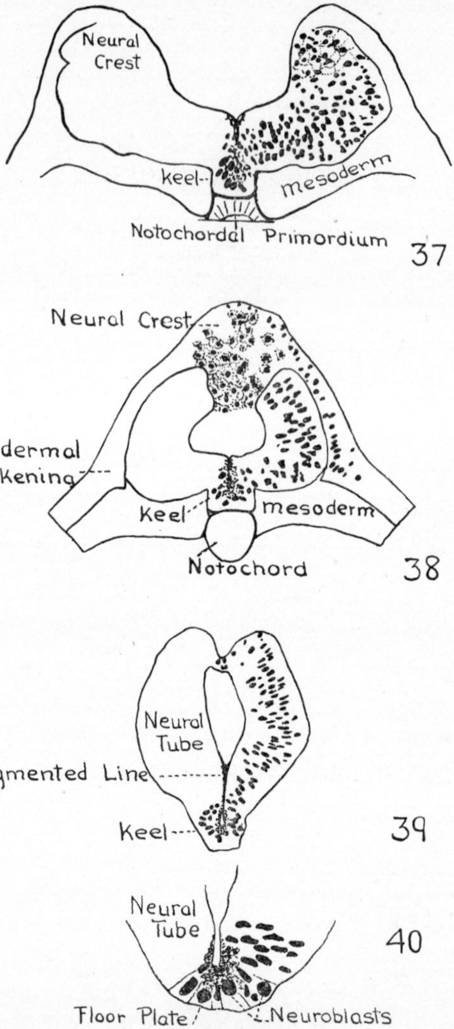

Figs. 37 to 40. Sections through the nervous system and subjacent structures of *Amblystoma* of a later stage of development than represented by Figs. 33 to 36. Copies respectively of Baker's Figs. 10, 11, 12, 14, with modification in labelling (2). The extension of the nervous system far dorsad above the mesoderm in Figs. 37 and 38 should be noted; also the formation of a cleft in the neural keel in Fig. 39 and the opening of this cleft into the dorsal part of the central canal in Fig. 40.

dorsal part of the nervous system is very remote from the centre of action of the mesoderm. Or, to state the condition in reference to nervous function, that part of the nervous system which is to become motor is located in immediate proximity to the mesodermal gradient, whereas that part which is to become sensory is far removed from the centre of action of the mesodermal gradient and subjected predominatingly to the ectodermal gradient.

HISTOLOGICAL FEATURES OF THE PARTS CONCERNED

During the shifting of gross relations as just described, important histological changes are occurring in the median part of the medullary plate. A shifting of cells takes place from a deeper position in the middle part of the medullary plate to near the ventral surface of the plate. This migration of cells, in fact, initiates the formation of the keel (Fig. 34). These cells shift "from the dorsal region of the neural plate and from parts immediately lateral to the mid-ventral and paraventral portions of the plate, resulting in the grouping or crowding of cells in this particular place. Simultaneously, a ventral downgrowth appears in the keel region of the neural plate." As the keel begins to form, pigment accumulates along the median longitudinal groove in the outer surface of the plate. As the keel advances in development, this collection of pigment extends deeper into the plate and another segregation of pigment occurs in the central part of the keel itself (Fig. 35). Also, as the keel reaches its highest degree of differentiation, its pigment collects to the median plane. As this occurs the cells composing the keel become radially arranged with their outer ends facing the mesoderm and notochord and their inner ends charged with pigment (Fig. 36).

Soon after the closure of the neural groove to form the neural tube, a cleft appears in the ventral part of the keel in the centre of the pigmented region (Fig. 39). This cleft is formed by the separation of the pigmented ends of the cells and extends more and more dorsad as development proceeds, until it finally becomes confluent with that part of the central canal which arises directly from the invagination of the medullary plate. At this time the differentiation of neuroepithelial cells to form nerve cells is far advanced in the neural keel (Fig. 40). In fact, such differentiation appears to have begun in the neural keel in the stage represented by Figs. 35 and 41, when the medullary plate is just beginning to fold upward along its margins.

PHYSICAL AND PHYSIOLOGICAL PROCESSES INVOLVED

The nerve cells that first differentiate out of the neuroepithelium of the neural keel become the motor neurones of the swimming mechanism. This being established, it is important to know something of the physical and physiological processes that are involved with the structural modification of this part of the nervous system in its transition from undifferentiated neuroepithelium to a specialized conducting mechanism.

During the early phase of this transition, as we have already observed, the mesoderm comes into close relation with only the lateral part of the medullary plate (Fig. 34). At this time, with the embryo opened from the ventral side and the entoderm removed, the continuous sheets of mesoderm may, with a fine dissecting needle, be turned back freely from the medial to a lateral position. But after the neural keel is formed, and the mesoderm, still unsegmented, has grown medially into contact with it, the neural keel and the mesoderm adhere

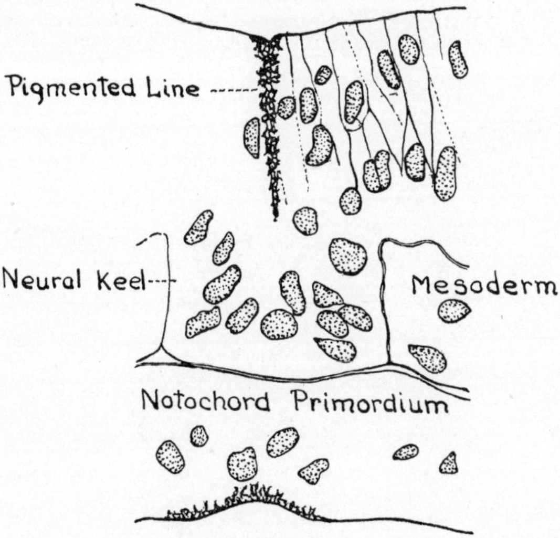

Fig. 41. Section through the middle portion of the neural plate, neural keel and entoderm at about the time that the mesoderm makes itself fast to the keel as described in the text. The beginning of a radial orientation of the epithelial nuclei of the keel is here apparent. In the middle of the keel, among the most ventral nuclei, is a large round nucleus which is in form and position characteristic of the earliest neuroblasts. Copy of Baker's Fig. 1, with modification in the labels (2).

to such a degree that the mesoderm cannot be dissected away as in the former instance without violent tearing. This demonstrates that there is a definite and specific physical reaction between the mesoderm and the neural keel.

Furthermore, simultaneously with the development of this adhesiveness between these two structures, the neuroepithelial cells of the region construct from their outer ends the external limiting membrane; and the cells that constitute the neural keel establish their radial orientation (Figs. 33 to 36). At the same time, also, pigment collects in the central end into dense masses. Soon after this has taken place, the central ends of those cells which face each other along the median plane of the nervous system separate in such a way as to form a cleft (Figs. 38, 39). This means that while the outer ends of the cells of the neural keel develop adhesiveness to adjacent structures their central ends lose that property. All of these changes are an expression of a specific physiological polarity in response to or in correlation with the changes in environmental conditions. Also, it is at this time that neuroepithelial cells begin to differentiate into neuroblasts in the neural keel (Figs. 35 to 41). Here, then, in the most medial part of the medullary plate, in immediate proximity to the most intimate contact of the nervous system with the chief mass of the mesoderm, is the first localized centre of differentiation of neuroepithelial cells into motor nerve cells. As just noted, the indifferent epithelial cells here become polarized in response to relations with the mesoderm. The same relations, then, may be a factor in the polarization of those cells which are differentiating into nerve cells. In view of this possibility, the process of transition from an indifferent cell to a nerve cell is an important consideration.

DIFFERENTIATION OF NEUROEPITHELIAL CELLS

When the cells of the neural tube of *Amblystoma* begin to differentiate into nerve cells they are organized in a single layer in radial arrangement. The differentiation of one of these cells into a nerve cell involves the massing of the protoplasm, together with the nucleus, near the external limiting membrane. That part next to the membrane becomes the growing tip of the cell. It spreads

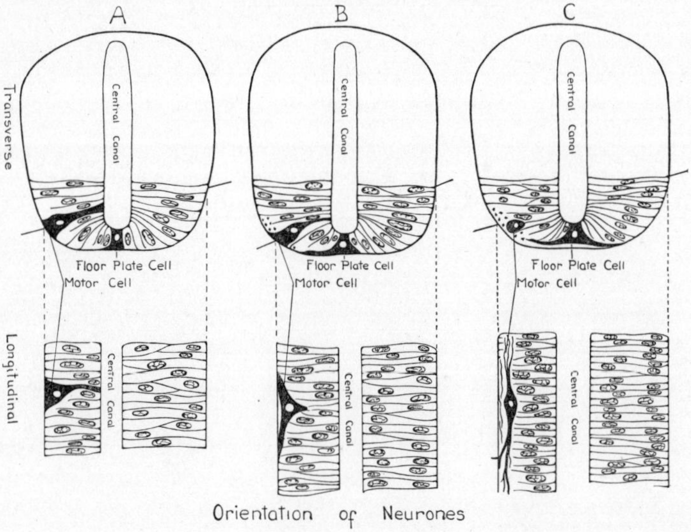

Fig. 42. Diagrams to illustrate the mode of differentiation of neuroepithelial cells into nerve cells. The upper series of figures represent sections transverse to the longitudinal axis of the spinal cord; the lower figures, sections parallel to the longitudinal axis and in planes indicated by the lines to which, in the upper figures, the dotted lines lead as projections of the boundaries of the lower figures. The cytoplasm of a differentiating cell concentrates against the external limiting membrane and spreads out along it, in opposite directions, usually one pole being considerably in advance of the other in its development. The cell, first breaking loose from the membrane near the nucleus, ultimately becomes entirely free from it.

out along the inner face of the membrane first as a single process going out in only one direction. This process is destined to become or to give rise to the axone (Fig. 42). Later the growing tip near the nucleus spreads in the opposite direction to form the dendrite. These two growing tips creep along the inner face of the external limiting membrane for a considerable distance and then the part of the cell nearest the nucleus breaks free from the membrane, while the two growing tips proceed in their course still adhering to the membrane. Eventually the ends also free themselves from the membrane, and the cell becomes a neurone. In the meantime the growing processes of the cell have been forcing their way against the resistance of the indifferent epithelial cells of the tube. The differentiating cell is in this respect, therefore, doing work. It is analogous to a germinating seed. It is to be regarded as the dynamic unit of the nervous system before it becomes nervous in function. The direction in which the intrinsic forces of the cell shall act, that is to say, its polarity, is the question that immediately concerns us at this point. The differentiation of the cell, like the germination of the seed, is one problem; the determination of its direction of growth is another. Our discussion, therefore, turns again to the question of polarity.

AXIAL GRADIENTS AND POLARITY OF LONGITUDINAL NEURONES

The primary motor cells of *Amblystoma* establish their polarity with their long axes in the longitudinal plane of the spinal cord at about the time that the embryo as a whole begins to elongate from the spherical condition of the egg. This is when the longitudinal or axial gradients become operative. In this process of polarization of the motor neuroblast, a neuroepithelial cell that faces the external limiting membrane along the line of adhesion between the neural tube and the mesoderm spreads out along the membrane with one bud tailward and the other headward. These two buds creep along the limiting membrane as an amoeba creeps along the substratum, or as neuroblasts *in vitro* creep along filaments of foreign substances (25). In the meantime the anterior part of the mesoderm organizes itself into mesodermal segments, and opposite the middle of each segment processes of motor cells that grow tailward form growing buds at right angles to their axes. These buds penetrate the limiting membrane and apply themselves immediately to the muscle cells. During this time the muscle is in adhesive contact with the limiting membrane of the nervous system where the nerve fibre first touches it (18).

The intimate relation that maintains between the mesoderm and the motor cells in this early period of growth deserves careful consideration. The mesoderm, as already emphasized, is adherent to one side of the external limiting membrane of the nervous system and the nerve cell is adherent to it on the other. It would seem that this physical intimacy must necessarily involve an influence of the metabolic gradient of the mesoderm along the membrane upon the nerve cell somewhat after the manner of an action current. At any rate, it is a fact that the axone of the motor nerve cell grows into the region of higher rate of the metabolic gradient of the mesoderm*.

* Child has suggested that the axone grows "down the gradient," that is to say, into regions of lower metabolic rate. While this does not seem to be true for *Amblystoma* in so far as the author has been able to determine, a sweeping generalization on the subject should await an exhaustive study of primary and secondary gradients in the

Some of these neurones, and probably all, become thus polarized while the mesoderm is in the unsegmented condition. With the reorganization of the central mass of the mesoderm into segments the primary gradient of the tissue must also become similarly modified into secondary gradients. There is, at any rate, evidence of a metabolic gradient in the muscle segment, for the nuclei of the muscle cells are most abundant in about the middle third of the segment, and pigment accumulates in the same region in massive bodies that lie usually if

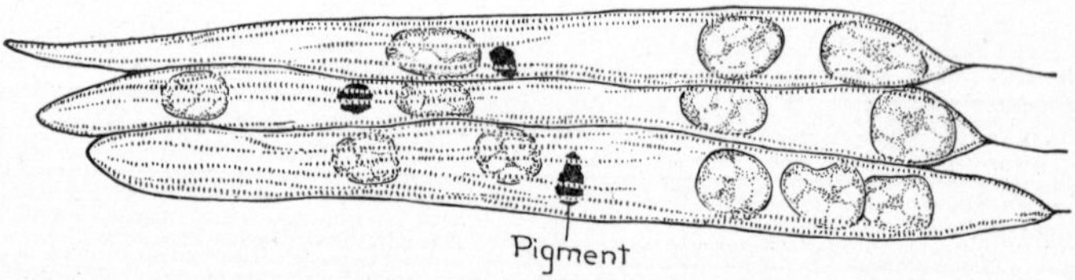

Fig. 43. A longitudinal section of muscle cells of the muscle segments of *Amblystoma punctatum* of the early swimming stage, to show the position of pigment masses as described in the text. The figures were made with the aid of the camera lucida.

not always alongside of nuclei (Fig. 43). Since it is well known that in the development of the amphibian egg pigment is formed most abundantly in the region of higher metabolic rate, the pigment of the muscle cell of *Amblystoma* may legitimately be regarded as an index of the relative rate of metabolism in the different parts of the cell. Interpretation upon this basis would make the

living tissues with reference to the polarization and growth of particular neurones that are adequately understood in their anatomy and development. It is hoped that such a study may soon be made of *Amblystoma*. In the meantime, Dr Child and the author are in agreement upon the more general principle that the orientation of neurones in growth is correlated with metabolic gradients.

middle of the muscle cell the region of high metabolic rate and the ends of the cell the regions of lower rate. The motor root branches of the descending axones of the motor cells would, therefore, be growing, like the axones themselves, into the regions of higher rate of a metabolic gradient, but of a local secondary gradient that has followed in the path of the primary gradient of the unsegmented mesoderm. According to this analysis, the inherent potentiality of growth of neuroepithelial cells and the dominance of mesodermal gradients over them, in activating polarity and directing growth, determine the primary pattern of the motor mechanism, which, as demonstrated in the first lecture, integrates the early behaviour of the animal.

Passing now to the consideration of the sensory mechanism, we recall that the primary sensory cells of *Amblystoma*, as in embryos of other aquatic vertebrates, lie within the spinal cord, and that they are of essentially the same type anatomically as the primary motor cells (Fig. 9). They are like the motor cells in that they form longitudinal conductors in the central nervous system and have side branches that form peripheral nerves (13). They differ from the motor cells in respect to the direction of conduction: the motor cells conduct tailward while the sensory cells conduct headward; and while the peripheral branch of the motor cell is an axone, that of the sensory cell is a dendrite. They differ from the motor cells, also, in position, and, therefore, in environmental conditions. While the motor cells develop in the median and ventral part of the neuroepithelium in immediate relation with the most concentrated mass of the mesoderm, as already explained, the sensory cells develop in the most lateral part of the medullary plate and at a relatively great distance from the centre of activity of the mesoderm; and as the margins of the medullary plate fold up to form the neural tube this remoteness of the sensory cells from the mesoderm is further accentuated, for the sensory cells come to lie in the dorsal part of the spinal cord while the mesoderm still does not rise above the ventral margins of the cord (Fig. 38).

The sensory cell, as in the case of the motor cell, forms a growing tip that faces the external limiting membrane. Also, this tip bifurcates and its two divisions creep in opposite directions along the external membrane, the one headward, the other tailward. Correlated with this orientation of the cell is the existence of the metabolic gradient of the nervous system which acts longitudinally, with the centre of high rate of metabolism headward (Figs. 30, 31). Since the headward growing process of the sensory cell becomes the axone, we find, then, sensory axones as well as motor axones growing into the region of high rate of metabolic gradient. Furthermore, when side branches grow off of the sensory cells to form the peripheral sensory nerves to the muscles, they grow to the end of the muscle cells. According to our earlier analysis, the end of the muscle cell is the region of low metabolic rate of the muscle. If these observations are correct, both sensory and motor axones grow into regions of high metabolic rate while dendrites of both sensory and motor cells grow into regions of low metabolic rate along axes of metabolic gradients.

But this statement requires qualification in order to represent correctly the determination of functional polarity in the nerve cell, for it has been demonstrated that the physiological polarity of these very sensory and motor neurones with which we are dealing is not a predetermined and fixed polarity. Segments of the embryonic spinal cord have been excised and replaced in the axis of the nervous system in such a way that the end of the piece that was nearest the head faces tailward (21, 29). In this reversed position topographically, the transplanted segments become part of a normally functioning spinal cord, with dorsal cells conducting headward and ventral cells conducting tailward. Therefore it is not exactly correct to say that axones grow into regions of higher rate of the physiological gradients, and dendrites into regions of lower rates, but the facts are that those processes of neuroepithelial cells which grow into regions of higher rate of a metabolic gradient become axones while those which grow into regions of lower rate of the gradient become dendrites.

It has been explained earlier in the lecture how a cleft forms in the centre of the neural keel of the medullary plate of *Amblystoma* to form ultimately the ventral part of the central canal of the spinal cord (Figs. 39, 40). The floor of this cleft becomes the ventral wall of the cavity and is known as the floor plate. Some of the cells of this plate form growing tips lateralward that creep along the external limiting membrane after the manner described for the motor and sensory cells (15). These growing tips are for a considerable period unipolar. They creep lateralward, or lateralward and tailward, toward the region of contact between the spinal cord and the mesoderm, and having reached this region turn caudad in the motor path. Later another growing tip forms at the base of the first and grows in the opposite direction. But, instead of joining the motor system, this tip branches out in immediate proximity to the floor plate to form with the axone terminals of sensory neurones the first neuropil of the nervous system. Since the commissural cells are confluent as a group with the cells of the motor tract, and since their axones grow diagonally into the motor tract, it is probable that their direction of growth, like that of the motor cells, is determined by the longitudinal gradient of the mesoderm. If this be true, then we may conclude that the mechanism that determines the primary behaviour pattern is organized under the dominance of preneural metabolic gradients.

DIFFERENTIATION IN LOCALIZED CENTRES

But these processes of differentiation and polarization of nerve cells do not take place evenly or at the same rate throughout the whole nervous system. They are greatly accelerated in particular centres of activity, and these centres conform to a definite pattern in progressive development (16).

Throughout the spinal cord and rhombencephalon differentiation goes on more rapidly in the ventral than in the dorsal region (Figs. 44, 45). This appears to be anticipated by the early differentiation of neuroblasts in the neural keel of the medullary plate. From there differentiation spreads outward and upward into the dorsal region. The primary sensory cells in the dorsal part of the spinal cord appear to violate this principle (13). But they differentiate precociously under some special *régime*, and are transitory. The other cells, which become definitive neurones, in their vicinity, lag far behind them in differentiation, and conform to the order of development from the mid-ventral region outward.

But there is not only localization of centres of acceleration in differentiation in the transverse plane. It occurs

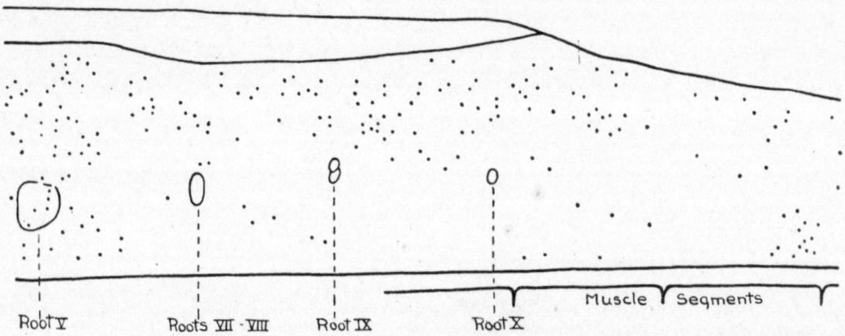

Fig. 44. An optical projection of the mitotic figures in the brain and spinal cord of *Amblystoma* of the coil stage upon the median plane, from the level of the root of the trigeminal nerve through the first three spinal segments. The segregation of the figures in a relatively dorsal position is noteworthy.

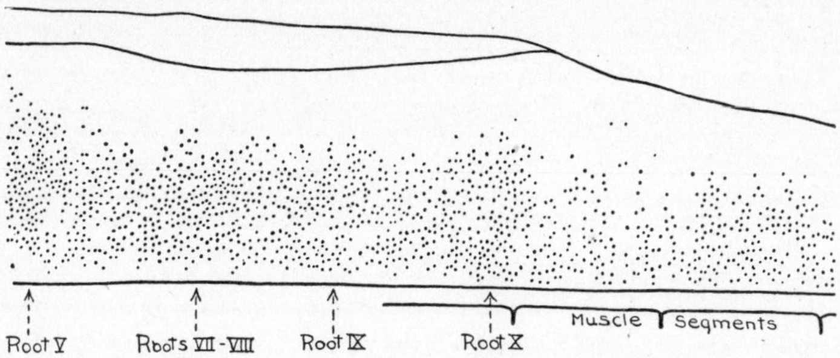

Fig. 45. An optical projection of the neuroblasts of the same part of the brain of the same animal as represented in Fig. 44. The segregation of the neuroblasts in a relatively ventral position is obvious.

also according to a definite pattern along the longitudinal axis of the nervous system (17). In that part of the spinal cord which lies behind about the sixth spinal segment, the cells that are undergoing differentiation are distributed quite evenly. This condition prevails at least until the early swimming stage. During this time there is no positive evidence of a metameric or other localization of such cells in this part of the cord. But in the more rostral part of the spinal cord and throughout the brain there prevails a definite pattern of localized centres in which differentiation is progressing at a much more rapid rate than in other parts (Fig. 46).

Two major centres of activity of this kind appear between the isthmus and the level of about the sixth muscle segment. One of these centres lies in front of the facial nerve root and the other behind it. These may be called prefacial and postfacial centres, as opposed to the still more caudal region of the spinal cord which may be designated as the spinal region; although, of course, a large portion of the spinal cord as understood morpho-

Legend to Fig. 46.

Fig. 46. Graphs representing the distribution of neuroblasts in the prefacial (from the isthmus to the level of the facial nerve root) and the postfacial (from the level of the facial nerve root through the sixth spinal segment) regions of the non-motile (*A*) and early flexure (*B*) stages of *Amblystoma punctatum*. The solid lines represent the motor system; the dotted line, the sensory system. The axis of ordinates indicates the number of cells occurring in successive 10 μ levels, as represented by the abscissae. Along the latter the successive vertical lines represent the isthmus (*F*), the levels of the several nerve roots (V, VII, IX, X) and the levels of myosepta. Graph *A* illustrates the fact that a definite pattern of both the sensory and motor systems is established before there is a conduction path from the receptor field to the muscle, and that the patterns of the two systems differ in important features, particularly with reference to the location of the centres of highest rate of differentiation. Graph *B*, representing the condition very soon after conduction from the receptors to the muscles begins, shows that the centre of greatest acceleration of differentiation in the sensory system is not adjacent to the root of a sensory nerve. It occurs, on the other hand, at the level in which the cerebellum develops. There is a strong suggestion of a similar centre even in the non-motile stage (*A*). In graph *B* the major centre of differentiation is not centred about the points of entrance of sensory nerve roots. It is opposite the auditory vesicle, and represents, in the neural tube condition, a centre that was directly adjacent to that part of the neural crest which gave rise to the auditory vesicle and related placodes. (See Figs. 5 and 6 of Landacre(34).)

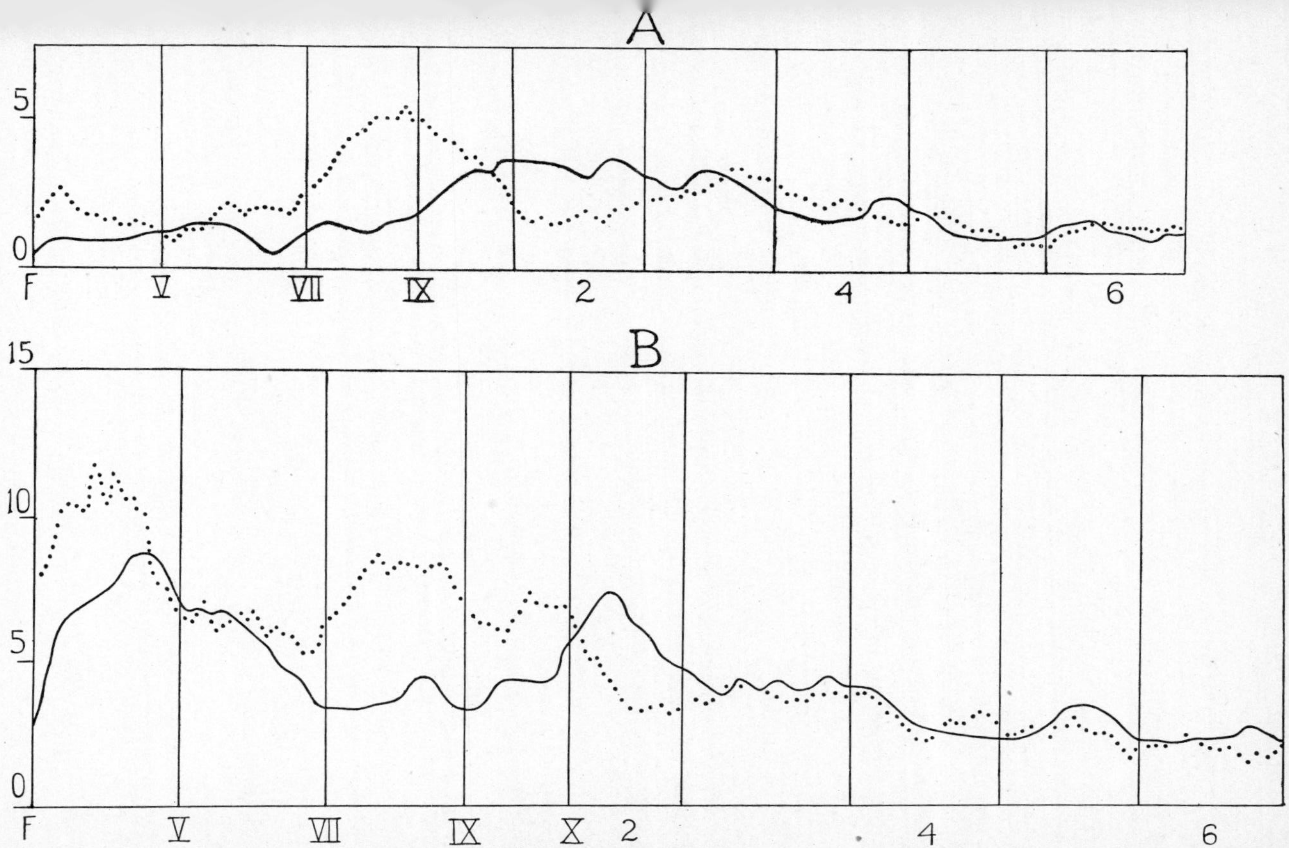

Fig. 46

logically lies within our postfacial region. Since the motor and sensory zones of these regions follow somewhat different patterns of differentiation it is necessary to deal with them separately.

Within the motor zone the prefacial and postfacial centres of acceleration are clearly defined in the non-motile stage of the *Amblystoma* embryo. At this time the postfacial centre involves particularly the first, second and third spinal segments without any sharp centralization. Within this major centre there is indication of secondary centres that appear to be metameric in arrangement. In the early flexure stage this postfacial centre is sharply centralized in the second spinal segment; but it still appears to express some degree of metamerism. In the coil stage there is evidence that the centre of greatest acceleration in this region is shifting headward; while in the early swimming stage it has become definitely localized in the first spinal segment.

Meanwhile the sensory system in the postfacial region appears in the non-motile and early flexure stages with its primary centre of acceleration in front of the first spinal segment and at the level of the auditory vesicle (19). In the coil stage this centre has shifted relatively caudad; whereas in the early swimming stage it, also, is definitely localized in the first spinal segment.

In the prefacial region, during this same period of development, there appears in the motor system, first, a centre of acceleration near the level of the entrance of the trigeminal nerve root. Later two centres arise in front of this, and the more anterior, by the early swimming stage, gains the ascendancy over the other two, which appear to become confluent with it. The sensory system of the prefacial region, on the other hand, has from the first its centre of greatest acceleration near the isthmus. Two other subordinate centres become more or less incorporated with it.

The motor and sensory systems, therefore, follow different patterns in their early periods of growth in both prefacial and postfacial regions. The motor system expresses metamerism in a minor way, but the major factor must have some other meaning than metamerism as it is ordinarily interpreted, for there is a shifting of the centre of acceleration with age. At first the centre of acceleration is localized only indefinitely in the region of the anterior part of the muscular (effector) mechanism. Later, as the centre of rapid differentiation develops in the sensory system in front of this, the motor centre of growth shifts forward towards the sensory and becomes sharply localized exactly at the anterior end of the muscle system. Meanwhile the primary centre of differentiation in the sensory system appears first opposite the region of ectoderm that is giving rise to the auditory vesicle and the ganglia from which the sensory nerves of that region grow out to the skin (34). Its primary localization is therefore related to the sensory (receptor) field. But as growth proceeds this centre apparently enters into a reciprocal activity with the centre of acceleration of the motor system in such a way that the two centres approach each other and come to occupy the same level in the nervous system. However, this shifting of position in the centres is not brought about by migration of cells. It is accomplished by a shifting of the centre of highest rate of differentiation. The primary centre of acceleration in the motor system is, then, associated with the dominating end of the muscle (effector) mechanism, while the primary centre of acceleration in the sensory system is associated with the dominant centres of the sensory (receptor) field in development.

In the prefacial region of the head there is no effector mechanism comparable to that in the postfacial region. At the time when the prefacial centres of acceleration in the differentiation of the brain become operative, the muscle-forming tissue that eventually gives rise in the related region to jaw muscles and eye muscles is in the condition of mesenchyme. The primordia of the nuclei of the eye muscles are in front of the isthmus and do not enter into this prefacial centre of activity. It may be that the two most posterior centres of acceleration in the motor zone of this region are concerned with the development of the motor nuclei of the trigeminal and facial nerves; but the most anterior centre lies too far forward for such a relation. It must be essentially related to something that is intrinsic in the brain. It appears to be an integral part of the primordium of the cerebellum.

The early differentiation of the sensory zone of the prefacial region appears to be primarily associated with that part of the neural crest which gives rise to the trigeminal ganglia (34). But acceleration of differentiation in the region immediately behind the isthmus is the beginning of the cerebellum. Furthermore it is apparently in correlation with this great activity in the sensory zone that the motor zone accelerates in differentiation in this region. It appears, therefore, that there is reciprocal influence between the sensory and motor zones in both the prefacial and postfacial regions of acceleration.

Within the cerebrum the motor system is a factor in differentiation only in the mesencephalon. The remainder of the brain in front of the isthmus must be regarded as sensory or associational in function. In these

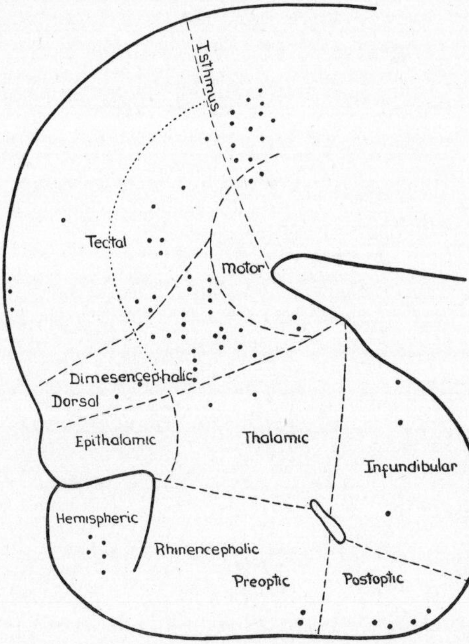

Fig. 47. Optical projection upon the sagittal plane of neuroblasts on one side of the midbrain and forebrain of an embryo of the non-motile stage of *Amblystoma punctatum*. The topographical regions named on the figure are determined according to the distribution of neuroblasts of this and older stages. The common boundaries of the dorsal and ventral sectors of the tectal and dimesencephalic regions is indicated by a light dotted line; the boundaries between major regions by the heavier broken lines.

parts, also, differentiation of the neuroepithelium into nerve cells proceeds from definite centres of acceleration (Fig. 47)(20). Three such centres make their appearance almost simultaneously. They are present in the non-motile stage of *Amblystoma*. One of these is located along the boundary between the mesencephalon and diencephalon, a second in the ventral region behind the chiasma ridge, the third in the cerebral hemisphere. The first of these we may call the dimesencephalic region; the second, the postoptic; the third, the hemispheric. Later similar centres arise in the tectal, thalamic, infundibular and olfactory regions. These centres soon become more or less confluent along contiguous territories, but the three primary centres maintain in general a dominance over the other centres, at least till the early swimming stage.

LOCALIZED CENTRES OF DIFFERENTIATION AND THE DEVELOPMENT OF CONDUCTION PATHS

Centres of differentiation in the brain and spinal cord arise as centres of disturbance in a placid lake. The disturbance, however, is of greater intensity or amplitude in some centres than in others. This difference appears

Legend to Fig. 48.

Fig. 48. Graphs to illustrate the relative rates of differentiation of neuroblasts in the several regions of the brain and in the several stages; non-motile stage, dotted line; early flexure, dash line; coil, dash-dot line; early swimming, solid line. In each stage the number of neuroblasts in the postfacial region is taken as 100 per cent., and the number of neuroblasts in the other regions for each age is stated in percentages of the number in the postfacial region of the corresponding age. In the non-motile stage the number of neuroblasts in the spinal region is 90 per cent. that of the postfacial, whereas in the other three stages the numbers are 55, 56 and 58 per cent. Or in the case of the cerebrum the number in the non-motile stage is 19 per cent.; in the early flexure, 28 per cent.; in the coil, 38 per cent.; in the early swimming stage, 70 per cent. This shows that the cerebrum as a whole is a centre of acceleration with reference to the postfacial region. In the last two stages the prefacial is accelerating with reference to the postfacial region. This graph takes no account of the actual increase in number of cells in the postfacial region from stage to stage. It illustrates only the relative accelerations and retardations of differentiation in the major divisions of the central nervous system. It is constructed upon the basis of the enumeration of cells in the right and left halves of 21 specimens. (For details of the method, see Coghill (17).)

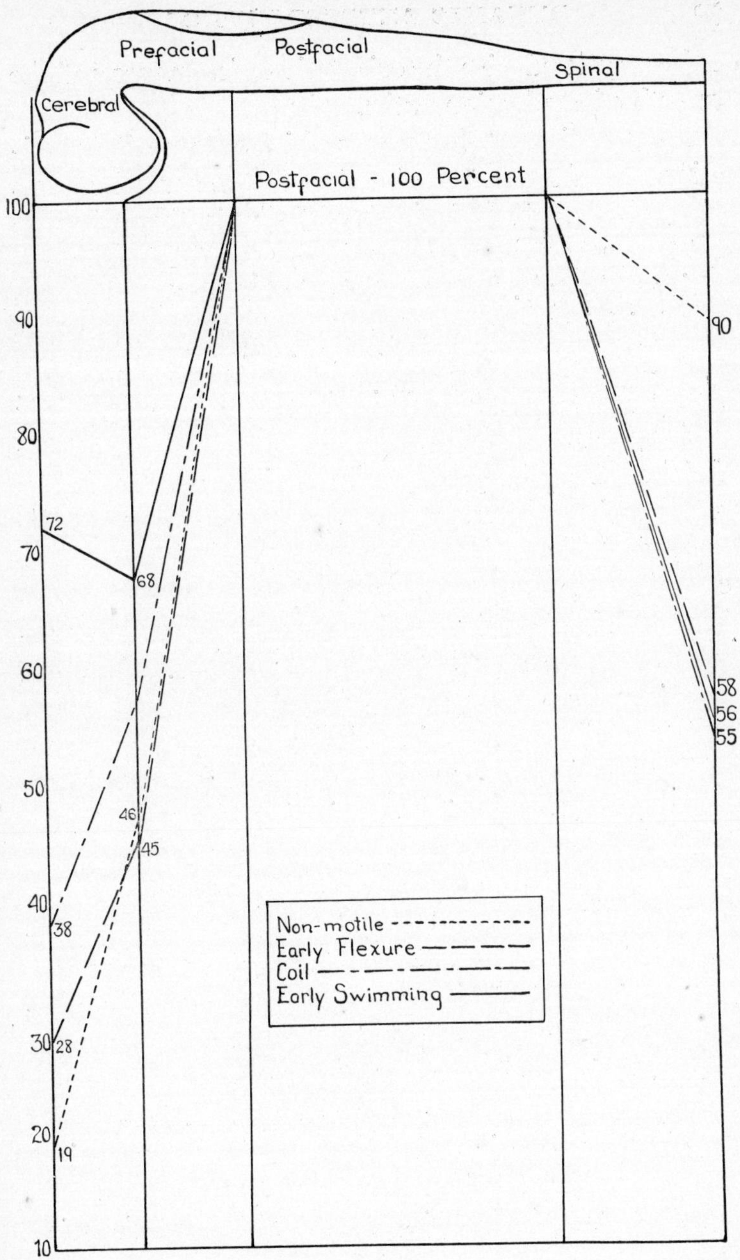

Prefacial
Postfacial
Spinal
Cerebral
Postfacial - 100 Percent
100
90
80
70
60
50
40
30
20
10
72
68
46
45
38
28
19
90
58
56
55
Non-motile
Early Flexure
Coil
Early Swimming

Fig. 48.

to be correlated with the orientation of neurones and the growth of conduction paths.

A comparison of the rates at which cells are differentiating out of the neuroepithelium in the several centres along the morphological axis of the brain shows that up till the early swimming stage the postfacial is the most active (Fig. 48). The prefacial is less active than the postfacial but more active than the dimesencephalic centre. Correlated with this relation is the fact that the first cells to become polarized in the dimesencephalic region grow caudad towards and eventually into the prefacial centre (Fig. 50). As differentiation begins in the tectal region, the cell processes grow towards the dimesencephalic and motor centres, and, in the more caudal parts, towards the prefacial centre. The first cells in the thalamic region to become polarized send processes towards the dimesencephalic centre. Similarly, cells in the dorsal part of the infundibular region grow first towards the thalamic and dimesencephalic regions, while those in the ventral part grow towards the postoptic region. The earliest polarization of the cells in the postoptic region is in the transverse plane, forming a commissure, just as in the case of the floor plate of the medulla oblongata and spinal cord. Later cells differentiating more laterally in this region send processes ventrad into the midventral region, and, finally, processes dorsad towards the thalamic and dimesencephalic region. This also is according to the manner of growth in the rhombencephalon and spinal cord. Very soon cells in the thalamic region become bipolar and orient themselves in general along lines between the postoptic and dimesencephalic regions. Meanwhile the cells which differentiated early in the dimesencephalic region have become bipolar with their processes growing forward towards the site of the posterior commissure. This, like the growth of cells in the prefacial or postfacial centres, is first ventrad and then dorsad. Axones of other cells also join these to form the posterior commissure in the early swimming stage.

In front of the optic chiasma, as already noted, the centre of highest rate of differentiation is in the cerebral hemisphere (Fig. 49). This centre of differentiation makes its first appearance near the region that is to become the primary olfactory centre. Cells that differentiate later more dorsally become polarized with their growing processes directed towards this earlier centre; and as cells differentiate in the olfactory centre, their first processes also grow towards the cerebral hemisphere. As the preoptic centre differentiates, its cells become first oriented with processes growing towards the primary olfactory centres. Later cells between these two regions become oriented with their longitudinal axes along lines between the primary olfactory and the preoptic region. Subsequently, between the coil and early swimming stages, cells in the preoptic region grow back over the base of the optic stalk into the infundibular region. This occurs as differentiation in the infundibular region becomes greatly accelerated.

Throughout the whole nervous system of *Amblystoma*, as we have seen, the differentiation of neuroepithelial cells into nerve cells begins in localized centres. From these centres differentiation spreads through the neuroepithelium, but, for a considerable time, the original centres maintain a greater activity than do the contiguous

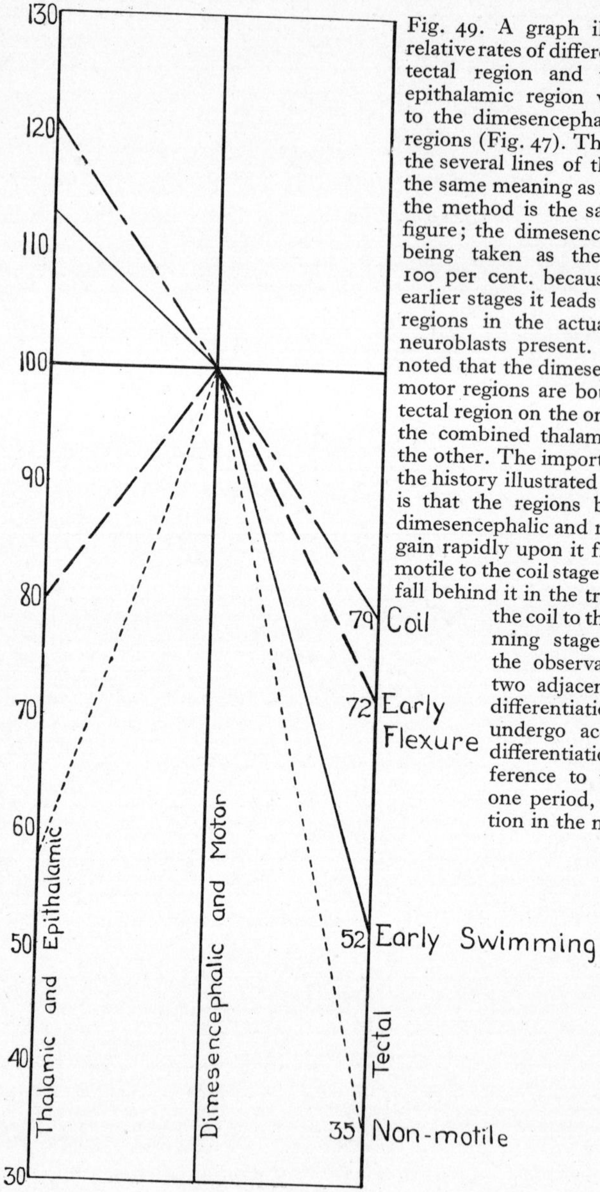

Fig. 49. A graph illustrating the relative rates of differentiation in the tectal region and thalamic plus epithalamic region with reference to the dimesencephalic and motor regions (Fig. 47). The structure of the several lines of the graph have the same meaning as in Fig. 48, and the method is the same as in that figure; the dimesencephalic region being taken as the standard of 100 per cent. because in the two earlier stages it leads the other two regions in the actual number of neuroblasts present. It should be noted that the dimesencephalic and motor regions are bounded by the tectal region on the one side and by the combined thalamic regions on the other. The important feature of the history illustrated by this graph is that the regions bordering the dimesencephalic and motor regions gain rapidly upon it from the non-motile to the coil stage and that they fall behind it in the transition from the coil to the early swimming stage; illustrating the observation that of two adjacent centres of differentiation one may undergo acceleration in differentiation with reference to the other in one period, and retardation in the next.

regions. While this condition prevails, cells which differentiate in relative proximity to a centre grow first towards that centre, and cells which lie between two such

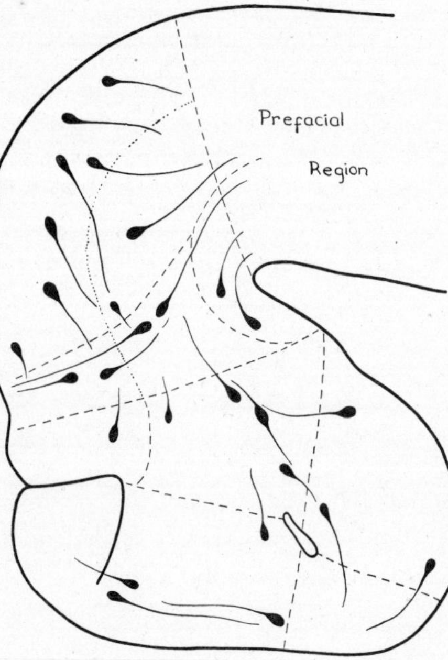

Fig. 50. Outline of the optical projection of the cerebrum and its regions according to Fig. 47, with the direction of growth of neuroblasts in the several regions schematically represented. Comparison with Fig. 47 will show that the cells first differentiating in the ventral part of the dimesencephalic region grow first towards the prefacial region, as do those in the more caudal part of the tectal region. Those farther forward in the tectal region and in the thalamic regions near the dimesencephalic grow toward the latter, which during the early periods is the most active centre (Figs. 47 and 49). Cells in the dorsal part of the infundibular region grow into the thalamic; those in the ventral part, into the postoptic centre, as do the cells in the ventral part of the thalamic.

centres of acceleration orient their growth along lines between these centres. Under the operation of this law we find many of the important conduction systems of the brain laid down. Among these are olfactory paths, the posterior commissure and its connection with the fasciculus longitudinalis medialis, the postoptic commissure, and fundamental paths between the thalamus and hypothalamus, and a path that is probably one of the forebrain bundles.

But there is still another feature of importance about the centres of acceleration in differentiation; they fluctuate in relative rates of activity (16, 20). Of two contiguous centres the one and then the other will have the higher rate of differentiation (Fig. 49). Since axones appear to grow into centres of higher rate, this fluctuation in relative rate may be the occasion for reciprocal conduction paths. As an example of this, the cerebral hemisphere has a higher rate of differentiation than the olfactory region in one period and a lower rate in the next. Reciprocal paths between these parts of the brain are well known. The essential conduction paths of the cerebrum, accordingly, as well as the primary neural mechanism, may be regarded as emerging from a preneural dynamic pattern.

LOCALIZED CENTRES OF DIFFERENTIATION AND SECONDARY GRADIENTS

Returning to the consideration of the medullary plate stage of the spinal cord and brain, we recall that the ventral side of the plate can be exposed to view, and that the middle part of it, which is thus clearly visible, known as the neural "keel," is the region in which differentiation of nerve cells is going on most rapidly. Allied with this process of differentiation is a condition that is of paramount importance in this discussion: this centre of highest rate of differentiation is at the same time the centre of greatest susceptibility to the action of potassium cyanide (2). If the medullary plate is flooded from the ventral side with an appropriate solution of potassium cyanide, the middle or "keel" portion disintegrates at such a rate that the plate cleaves along the middle line while its lateral portions still show no signs of disintegration (Figs. 35 and 41). In this instance, therefore, a centre of accelerated differentiation is coincident with the region of high rate in a metabolic gradient*. Furthermore, as differentiation progresses lateralward in the plate and dorsalward in the neural tube, the first process of every nerve cell to grow in the transverse plane grows towards this centre of differentiation, now known to be also the high region of a metabolic gradient.

In this demonstration of the coincidence of a centre of differentiation with the high region of a metabolic gradient we have grounds for a working hypothesis that all centres of differentiation as we have described them in the brain and spinal cord have the properties of

* Bodine and his associates find in their study of the pupae of insects that during the initial period of histolysis (dedifferentiation) there is decrease in oxygen consumption per unit of body weight, whereas there is progressive increase during the later period of differentiation of pupal tissue (3). They have not yet analysed in this respect the process of growth as regards proliferation of cells and differentiation of tissues, but it would seem that proliferation would be on the decrease with advancing age and increased differentiation. If this inference is true, that increase of differentiation accounts for the increase in oxygen consumption in the organism at large, then it would be expected that centres in the organism that have higher rates of differentiation would be also centres of higher rate of oxygen consumption as compared with other regions. Upon this hypothesis, metabolic gradients would be expected in coincidence with the localized centres of differentiation that have been described above as occurring in the nervous system of *Amblystoma*.

metabolic gradients, and that conduction paths emerge from these secondary gradients after the manner in which the primary longitudinal paths emerge from the primary longitudinal gradients of the mesoderm and ectoderm.

STATEMENT OF THE PROBLEM

The burden of this lecture is to place an old problem on a new basis and to present a working hypothesis for its solution. How do conduction paths of the central nervous system come to be where they are, or how do they acquire their definitive function?

In presenting this problem the lecture has dealt with primary conduction paths that take form before definitive physiological excitation can be regarded as playing any part in activating or directing the growth of nerve cells or fibres*. These paths must therefore rise out of activities of the tissue that are preneural, according to the conventional idea of nervous function. Also the differentiation of neuroepithelial cells into nerve cells occurs according to a preneural pattern of localized centres; and nascent nerve cells orient themselves definitely with reference to these centres. In this orientation the growing tips of the cells adhere to the external limiting membrane in such a way as to indicate

* Bok's idea of stimulogenous outgrowth of the axis cylinder, upon which Äriens Kappers largely builds his theory of neurobiotaxis, assumes a pre-existing conducting path to which polarization and outgrowth of axones and dendrites are tropic reactions (1, 4). This lecture is dealing especially with the origin of those primary conduction paths during a period which antedates definitive nervous impulses. The hypothesis offered here should be regarded as supplementing rather than refuting "neurobiotaxis."

that this membrane is an essential factor in directing their growth. One is, therefore, led to suspect that some of the physical processes that are characteristic of membranes are involved in the problem of polarity of the nerve cells and the origin of the conduction paths. Meanwhile, one centre into which nascent nerve cells grow has been found to be coincident with a metabolic gradient, and metabolic gradients are known to be, in part, gradients of electropotential. It is a reasonable hypothesis, therefore, that a gradient of electric potential acting along the external limiting membrane is the essential factor in determining the path along which the nascent nerve cell shall grow.

Whatever the merit of this hypothesis may be, it is at least supported by a new line of evidence that opens an approach to the problem through modern methods of general physiology and biophysics; for *Amblystoma*, which has given us the anatomical basis of the problem in adequate detail, is also an ideal animal for the application of those experimental methods which are required to determine the exact nature of the dynamic antecedents of neural mechanisms.

OVERLAP OF PRENEURAL UPON NEURAL PROCESSES

Up to this point in our discussion we have been thinking of the preneural integrating forces of the individual as antedating the neural functions; but, while that is true for the various particular conduction paths, it is not true for the nervous system as a whole after the first conduction path has become established. The longitudinal motor and sensory tracts, if our interpretation is correct, emerge from antecedent longitudinal metabolic gradients; but after these tracts have acquired nervous function the preneural processes are operating in other parts of the brain and spinal cord, neuro-epithelial cells are differentiating into nerve cells, centres of differentiation are arising, and nerve cells are orienting themselves with reference to these centres. All of this is preneural activity. The preneural system of integration, therefore, overlaps the neural in the regular course of development. And not only is this true for the nervous system as a whole, but it is true for the individual neurone, for, as demonstrated in the first lecture, motor neurones grow in purely embryonic fashion for a relatively long time after they become functional conductors in an integrating mechanism.

The significance of the overlapping of the preneural upon the neural processes constitutes the theme of the next and final lecture of the series.

Lecture Three

GROWTH OF THE NERVE CELL AND THE INTERPRETATION OF BEHAVIOUR

THE first lecture dealt with the nervous system as a conducting mechanism that determines the immediate behaviour pattern: in the second lecture the origin of the conducting mechanism from preneural structures and functions was outlined, with emphasis, in conclusion, upon the fact that the preneural functions overlap the neural functions in the development of the mechanism of behaviour. It is apparent from this fact of overlapping of preneural and neural processes that the conventional idea of conduction does not give us a complete picture of how the nervous system plays its rôle in the development of behaviour. It is the purpose of this lecture to show how the preneural process of growth and differentiation in the nervous system, and particularly in the individual neurone, participates in the function of the nervous system as a whole.

LEARNING AS DEVELOPMENT OF BEHAVIOUR

Our specific data on the subject are drawn from relatively early periods of the life of the animal when behaviour is in a process of perfectly obvious development. But in reality behaviour is always in process of development in animals that can learn by experience. When new turns in behaviour cease to appear in the life of the individual its behaviour ceases to be intelligent.

Stereotyped behaviour in its most intensive form characterizes those animals in which no special provision is made for the growth of neurones after the nervous system begins to function in determining the behaviour of the individual. In insects, for example, when they emerge from the chrysalis, the functional nervous system is not embedded in a matrix of embryonic nervous tissue as it is in the young of *Amblystoma* and other vertebrates. Insects, accordingly, learn relatively little from experience, whereas the highest degree of modifiability of behaviour is possessed by those

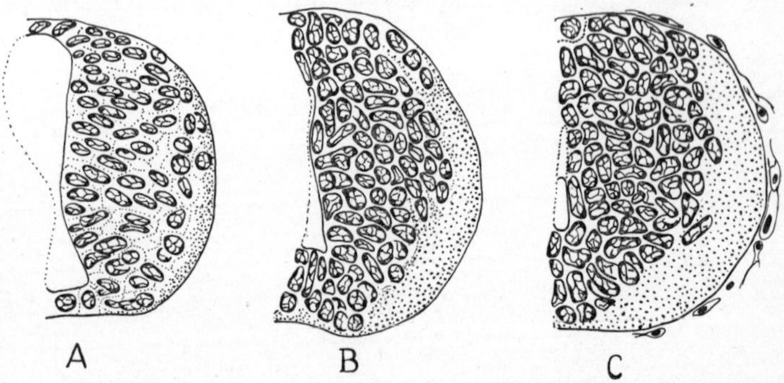

Fig. 51. Camera lucida drawings at the same magnification of transverse sections of the spinal cord at the level of the fourth spinal motor root of *Amblystoma punctatum*: *A*, of the early swimming stage; *B*, at the time when the fore limb moves only with trunk movement; *C*, in the early fore limb reflex stage.

vertebrates whose brain cells have the largest potentiality of growth when the nervous system begins to function as a conducting mechanism. Figs. 51 and 52 illustrate crudely the growth of the nervous system of *Amblystoma* from the time it begins to swim till it begins to walk. This potentiality of growth in the functional neurone of *Amblystoma* has a very definite and specific relation to the development of behaviour. We infer, therefore, that potentiality of growth is a factor in the development of behaviour of other vertebrates in proportion to the degree or scope of its occurrence; and that it is in the same proportion a factor in the process of learning, provided learning in its broadest sense connotes development of behaviour.

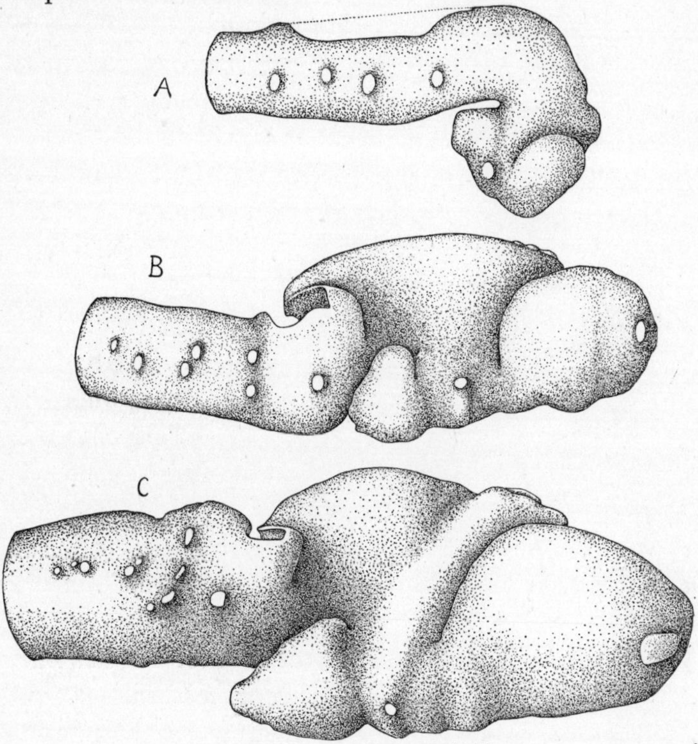

Fig. 52. Lateral views of models of the brain of *Amblystoma punctatum* made at a magnification of 200 diameters: *A*, at the early swimming stage; *B*, at the age when the limb can move only as the trunk moves; *C*, at the time when the animal begins to walk. The figures ×40.

INADEQUACY OF THE NEURONE CONCEPT

The neurone theory, as it is usually employed to elucidate the anatomy and physiology of the nervous system, deals with the nerve cell, at least by implication, as a definite and fixed unit of structure and function. It is commonly recognized, of course, that nerve cells increase their dimensions after they begin to function as conductors, as, for example, in the lengthening of the nerves in the limbs with the lengthening of the limbs in growth; and that upon severing the fibres of the peripheral system they will regenerate by processes of growth that are essentially embryonic in nature. But it is not generally recognized that after the nerve cell has assumed a definite and specific rôle as conductor in a functional mechanism it grows for a relatively long time in strictly embryonic fashion and thereby extends its sphere of action enormously.

Ramón y Cajal, probably more than any other investigator, has emphasized this phase of activity of the neurone, and he has in a very general way applied it in the interpretation of physiological and psychological phenomena (40). All of his discussions of the subject, however, appear to be based upon the hypothesis that the growth which occurs in neurones after the embryonic period, as the latter is ordinarily understood, is activated by exercise or intensified function in those parts of the nervous system in which it occurs. He has made no correlation of particular growth processes in neurones with specific results in behaviour. In fact, adequate data on such correlations have not formerly been available upon which to base hypotheses. On the other hand, our investigations, as presented in the earlier lectures, have revealed certain facts concerning the growth of the neurone and the development of behaviour that enable us to see the growth of the neurone and the development of behaviour in their actual relation to each other under definite and specific conditions.

Nerve cells, like seeds planted by a gardener, spring up and grow according to a definite pattern*. This pattern, in the nervous system, is established in its main outlines before nervous function, excitation or exercise begins (Figs. 46, 47). It is established, also, before vascularization of the nervous system occurs, which Ramón y Cajal holds is an intermediary process between increased nervous function and the growth of neurones under adult conditions. Nervous function or exercise, therefore, cannot be either directly or indirectly the cause of growth and differentiation of the nerve cell. This is confirmed by the growth of nerve cells *in vitro* in normal embryonic fashion (25). Under these artificial conditions their orientation and direction of growth can be determined by extrinsic agencies (30). Under natural conditions, also, the path along which they grow seems to be determined by surrounding conditions. Among these conditions, in early periods, as presented in the last lecture, are metabolic gradients of the organism and localized centres of growth or differentiation of nerve cells themselves. During later periods processes of conduction in established tracts may be one activating factor in directing the line of growth (1, 4); vascularization may be another (40). But, in the nerve cell as in the seed, growth as such must be regarded as the expression of an intrinsic potentiality of the cell.

Furthermore, this potentiality, as shown in the last lecture, is not exhausted when the nerve cell begins to conduct impulses according to its definitive rôle. After

* Kingsbury has used the figure of the germinating seed to elucidate his interpretation of cephalocaudal development of the nervous system (32).

this the cell continues to grow and seek out new realms to conquer. The functional nerve cell is, then, more than a conducting mechanism, whatever may be involved in conduction. It is from the beginning a dynamic system reacting to its environment after the manner of a living organism. Physiological conduction is, so to speak, its accessory or secondary function. If it ever loses its potentiality of growth and differentiation we do not know when or where*.

It is not proposed, of course, that nerve cells of higher vertebrates carry over into the later life of the individual the potentiality to grow long fibres within the central nervous system. The lack of evidence for the growth of new conduction paths in the brain of man or other mammals, even by way of regeneration after lesions, is fully recognized. But growth of such magnitude is not in mind when we speak of the potentiality, for example, of the cells of the cerebral cortex to grow after the brain has attained its apparently full growth in mass; nor must growth be so extensive in order to affect behaviour. Growth of the terminals of axones and dendrites through microscopic dimensions is sufficient to have profound effect in behaviour. This we have demonstrated in the first lecture in a vertebrate of such primitive form as *Amblystoma* (Fig. 11), which, by the growth of terminals

* That Purkinje cells enter upon retrogressive development relatively early in the life of the individual has been shown by Ellis for man and by Inukai for the rat (23, 31). These results, however, do not imply that other cells in the cerebellum may not meanwhile continue in a progressive development. An analogy of this is seen in the loss of Rohon-Beard cells or lateral line components of the cranial nerves in certain forms while other components of the nervous system continue in development.

of nerve cells over a distance of less than one one-hundredth of a millimetre, transforms itself from an animal that must lie helpless where chance places it into one that can explore its environment in response to impulses from within or stimulation from without*. This is for *Amblystoma* a discovery, so to speak, of incalculable significance. It is to all intents and purposes a solution of one of the crucial problems of life. Furthermore, as illustrated in the first lecture, this solution of the problem of transportation by water is applied by *Amblystoma* to the problem of transportation over land.

Psychologists may object to this connotation of problem-solving; but our treatment of the subject is biological. In the biological sense *Amblystoma* is certainly solving its immediate problems of survival by means of the growth of neurones that are already functional conductors. Such an interpretation can scarcely be rejected upon *a priori* grounds by the behaviourist who believes that the mechanism is the individual. It is not our purpose to question this belief. Our purpose, on the other hand, is to determine what is the real nature of the mechanism, and how it works. In so far as our knowledge goes concerning specific relations between the development of the nervous mechanism and the development of behaviour, the conception that a neurone grows during a certain so-called embryonic period, or period of maturation, and then ceases to grow and becomes simply a conductor in a fixed mechanism is

* Tracy has shown that the early behaviour pattern of the toadfish is established before the afferent system is functional and that the motor system is subject to endogenous stimulation, in part at least, by carbon dioxide (42). Similar excitation may occur in *Amblystoma*, but masked by a dominating afferent system.

erroneous, and wholly inadequate to account for the function of the nervous system as a mechanism of learning.

THE FORM OF THE BEHAVIOUR PATTERN VERSUS THE CONDITIONING OF PERFORMANCE

The relation that is found to exist between the structural development of the nervous system and the development of the ability of the organism to perform particular acts makes it necessary, in the present state of our knowledge, to distinguish sharply between the process by which the animal determines what acts it can do and the process by which it determines when and to what extent it will do them. That a muscle contracts, and in its contraction produces a particular form of activity, depends largely, if not wholly, upon developmental mechanics to which the value of the act in terms of definitive behaviour can have no immediate causal relation. The neuromuscular structural relations that make the act possible must exist before the act is performed.

A specific example of this may be drawn from the first lecture. It was there shown that a nervous motor mechanism is established in *Amblystoma* for some time before the animal responds to stimulation, that this mechanism is such as to conduct impulses to the muscles from the head tailward, and that this order of conduction to the muscles gives the resulting movement locomotor value and thereby becomes the basic principle of both aquatic and terrestrial locomotion. The general pattern of the primary nervous mechanism of walking is, therefore, laid down before the animal can in the least respond to its environment. Also, the conversion of the simple flexure into a compound flexure is brought about by the growth of side branches from certain nerve fibres into specific relation with other neurones; but swimming is the end result of this mechanical adjustment, not its cause. So it can be shown, as in the first lecture, that the form of the behaviour pattern in *Amblystoma* develops step by step according to the order of growth in particular parts of the nervous mechanism.

Accordingly, the normal experience of the animal with reference to the outside world appears to have nothing specifically to do with the determination of the form into which the behaviour of the animal is cast. On the other hand, experience has much to do with determining when and to what extent the potentiality of behaviour shall rise into action. This will be discussed presently.

"GESTALT" IN DETERMINING THE FORM OF BEHAVIOUR

The "Gestalt" school of psychology stands for total unity as the dominant principle governing mental processes (33). It seems, however, to have been concerned wholly with the processes that condition behaviour, and to have entirely neglected the processes which determine the form of the behaviour pattern. According to "Gestalt," a simple, pure or elementary sensation does not exist as such. There are no such units which combine to form perceptions. The perception is a "quality upon a ground": a total unity from the first. The apparently particular elements in consciousness emerge from a general field and exist only in relation to that field. This is equivalent, in the motor phase of the organism, to a totally integrated pattern in which partial patterns become more or less individuated. This principle was demonstrated in the first lecture as operating in the development of the form of the behaviour pattern of *Amblystoma*, and in the second lecture it was demonstrated for the origin of the mechanism which determines the early behaviour forms.

According to our interpretation of the observations presented in the second lecture, there is no hiatus between the preneural and the neural modes of integration. The longitudinally conducting motor path, for example, arises from cells that are under the dominance of a metabolic gradient which determines physiological polarity in the developing conduction path. The specialized conduction path emerges, so to speak, from the organismic gradient. It was demonstrated, also, how the general pattern of conduction paths in the midbrain and forebrain makes its appearance along lines of action of secondary metabolic gradients. Conduction paths do not come into existence, then, as absolutely new and discrete entities. They arise from a general field of organic activity by a process of specialization, emergence or individuation.

From the facts related in the first lecture it is obvious that there are two processes that are operating simultaneously in the development of behaviour. The one is expansion of the total pattern as a perfectly integrated unit; the other is the individuation of partial systems which eventually acquire more or less discreteness. In *Amblystoma* the total pattern first extends through the trunk and tail. As this pattern enlarges, the parts involved are always perfectly integrated. This totally integrated pattern then extends into the gills, next into the fore limbs and finally into the hind limbs. But as the totally integrated pattern expands through the organism, its parts, one after another, in the same order as they were invaded by the total pattern, begin to acquire a measure of individuality of their own: first the gills, then the fore limbs and finally the hind limbs. This means that local reflexes emerge as, in the language of "Gestalt," a "quality upon a ground"; that is to say, they emerge as a special feature within a more diffuse but dominant mechanism of integration of the whole organism. They cannot be regarded as simply the action of a chain of neurones, excepting as every link of the chain is conceived to be welded into the organism as a whole.

This principle is thoroughly demonstrated for *Amblystoma*, a typical vertebrate, and there is nothing in our knowledge of the development of behaviour to indicate that the principle does not prevail universally in vertebrates, including man. There is no direct evidence for the hypothesis that behaviour, in so far as the form of the pattern is concerned, is simply a combination or co-ordination of reflexes. On the contrary, there is conclusive evidence of a dominant organic unity from the beginning. That evidence appears not only in the manner in which behaviour develops, but particularly in the manner in which the nervous system puts the principle into effect, for, as shown in the first lecture, the nervous system concerns itself first with the maintenance of the integrity of the individual, and only later makes provision for local reflexes*.

* Tracy's studies on the development of behaviour in the toadfish (*Opsanus tau*) show that the earliest movements are localized and unco-ordinated (42). These are probably myogenic; for Tracy explicitly

GROWTH OF FUNCTIONAL NEURONES AND THE INTEGRITY OF THE INDIVIDUAL

The development of behaviour primarily through the extension of the total pattern, rather than through the projection of primarily isolated parts to become integrated secondarily, means that the maintenance of the integrity of the individual as a whole is the elementary function of the nervous system. This function is performed in *Amblystoma* through the growth of functional neurones into nascent organs. The same neurones, for example, that integrate the muscles of the trunk, while performing this function, grow into the limb by means of new branches long before the limb has power of movement. In like manner, the tissues of the tongue

holds that reflexes are not the primary components of behaviour. Although the toadfish is a very highly specialized member of the most specialized group of fishes (Teleostei), the writer finds nothing in the development of the toadfish, as described by Tracy, that is fundamentally inconsistent with the development of *Amblystoma*.

The work of Sherrington and his associates has, by their analytical method of study, brought the reflex out into such bold relief that its discreteness, as related to the behaviour pattern as a whole, has come to be overestimated and overemphasized by behaviourists who are not themselves technically familiar with the experimental results in a broad way (41). The synthetic method of Pavlov, also, has been accepted by many as exalting the place of discrete reflex mechanisms in behaviour, whereas the conditioned reflex involves only the afferent side of the arc and evokes nothing new in effector functions (39). The work of Magnus, on the other hand, emphasizes the solidarity of the total mechanism and the inseparable linkage of reflexes with it (37). This dominance of the total pattern over all of its components, virtually as Magnus saw it experimentally, is seen unmistakably in the development of the behaviour of *Amblystoma*. It is also obvious in the development of movements in the human foetus as recorded by M. Minkowski in numerous contributions which cannot be given appropriate treatment here. They will receive special attention in later publications.

receive branches from motor neurones that are engaged in integrating the trunk long before the tongue has muscle tissue in it. It is therefore the potentiality of the functional neurone to grow in embryonic fashion that gives to the organism as a whole its ability to subjugate new parts and thereby maintain its unity during the development of behaviour. Such growth of the already conducting neurones accomplishes, then, the primary function of the nervous system: the maintenance of the integrity of the individual while the behaviour pattern expands.

THE MECHANISM OF INDIVIDUATION

That part of the nervous system of *Amblystoma* which effects and maintains the integrity of the total pattern as it expands, for instance, into the limbs, is small as compared with the remainder of the mechanism of functional conductors in the brain and spinal cord. There is at this time a surplus of neural mechanism over and above that which is actually engaged in executing the immediate behaviour pattern. Some of this nervous surplus can be recognized as involving that part of the motor mechanism which eventually participates in the control of the movements of the limbs, the tongue and the extrinsic muscles of the eyes. It appears, therefore, that the individuation of a partial pattern or local reflex within the total pattern is anticipated in the central nervous system by the growth of a nervous organization with specific reference to that partial pattern long before the latter makes its appearance in behaviour. The details of this process are still to be determined; but its general nature is obvious.

"FORWARD REFERENCE" IN THE CONDITIONING MECHANISM

In *Amblystoma* the primary and secondary olfactory centres are relatively advanced in differentiation before the olfactory nerve enters the brain; conduction paths from optic centres to the brain stem, notably the fasciculus longitudinalis medialis, can be recognized before the optic nerve approaches the brain; and the fibres of the optic nerve penetrate the brain before bipolar cells or rods and cones can be recognized in the retina (14). The order of development of the conditioning system is, therefore, from the centre to the sense organ; and consequently there is in the association system as well as in the motor system of *Amblystoma* an overgrowth of neural mechanisms beyond the capacity of the animal to express their full nervous potential in behaviour.

Comparative embryological studies indicate that the higher the animal in the order of intelligence the more the neural overgrowth, as regards the immediate possibilities of behaviour, involves the conditioning mechanism; that is to say, the mechanisms that determine the time and degree of the performance as opposed to the mechanism that determines the form of the behaviour.

In *Amblystoma*, at the time muscular movements make their appearance, as described in the second lecture, nerve cells are only beginning to appear in the cerebral hemisphere and primary olfactory and optic centres, whereas in the human fetus of eighteen weeks the Betz cells of the motor cortex are highly differentiated, the motor and visual cortical areas are well defined by their characteristic lamination, and lamination has begun even in the prefrontal region (6, 20). This means that in man at a stage in development when body movements are of the simplest order, that part of the mechanism of association which deals particularly with the highest mental and moral processes not only is relatively massive but has definitely begun to organize itself into the mechanical pattern that characterizes it in the adult.

It appears, therefore, that the greater the possibility of behaviour in the animal the more the central conditioning mechanisms run ahead of the motor or effector mechanism in development. But in the motor system, by which the form of the behaviour pattern is determined, as described in the first lecture, this overgrowth of nervous organization beyond the immediate capacity of the animal to express the nervous potential, has specific reference to particular behaviour forms of the future. From this fact the inference is naturally drawn that the overgrowth which occurs in the mechanism that conditions behaviour has reference to conditioning processes of the future.

As just noted, while man's behaviour pattern is still of the simplest order, long before birth, his mechanism that has to do with his most refined adjustments in life has already the major structural features that characterize it in the adult (6). This overgrowth of the conditioning mechanism cannot intrinsically anticipate particular remote situations; but it must represent potentialities of behaviour that can come to full expression only in the future. This is not to say that the mechanism in question has no function in earlier periods, as we shall see later. But it certainly means that there is a mechanistic equivalent for man's ability to develop attitudes that can come to expression only in future behaviour. There is therefore, in man, "forward reference" in neural mechanism as well as in behaviour*.

DIFFERENTIATION IN THE CEREBRAL CORTEX

This interpretation of the human cerebral cortex upon the basis of what occurs in the development of *Amblystoma* might seem strained if it were not for certain facts that have recently been brought to light upon the correlation of structure and function in the development of the cerebral cortex of mammals. Langworthy has explored the cerebral cortex of new-born kittens with electric stimulation and has found that the centre which controls the fore limb is functional within a few hours after birth (35). This functional area involves the cruciate gyrus in front of the cruciate sulcus. Maturation, Langworthy says, "appears to begin on the surface of the gyri, gradually progressing toward the sulcus.... The cells lying in the depth of and on the anterior and posterior walls of the sulcus are the last to mature and throughout life this portion of the motor cortex appears less differentiated than the areas upon the surface." The work of Bolton and Moyes on an 18-weeks' human fetus makes it practically certain that the same mode of differentiation prevails in the human cerebral cortex (6).

These descriptions of progressive differentiation in the mammalian cerebral cortex give a good picture of the mode of differentiation in the brain of *Amblystoma*.

* Herrick develops the conception of the ability of man to formulate his behaviour with reference to future events or interests and calls such an act "forward reference" (26, 27).

Differentiation in both instances proceeds outward from localized centres in which nerve cells are not only most numerous but also from the beginning most highly differentiated. The mechanisms of association of *Amblystoma* and of mammals certainly follow the same law of differentiation from localized centres. A complete picture of the function of the cerebral cortex of mammals must, therefore, include preneural dynamic systems such as were considered in the second lecture as operating in the nervous mechanism of *Amblystoma*.

PROGRESSIVE CONSOLIDATION OF THE ASSOCIATIONAL INTO THE MOTOR MECHANISMS

In considering the mechanism of association one must guard against the use of terms as if there were fixed categories of function in the nervous system, such as sensory, associational and motor. Fixed categories are dangerous assumptions anywhere in science, and particularly so in dealing with the vertebrate nervous system, as we shall see at once.

In the earliest periods of differentiation of neuroblasts in that part of the brain which is categorically called sensory, no sooner has a neuroblast clearly differentiated from the neuroepithelium than it begins to creep along the external limiting membrane towards the motor system; and as development proceeds it grows more and more into the motor relation (19). A conspicuous case of this is the growth of neurones from sensory centres into the fasciculus longitudinalis medialis, which is primarily motor. These categorically sensory neurones become consolidated with the motor neurones to form a path that is co-ordinating or integrating in function (18).

This process of consolidation is the prototype of what transpires in the cerebral cortex when the Betz cells grow downward from a centre that is morphologically sensory or associational into the motor system to become consolidated into the mechanism of integration. In this instance the progressive nature of the process of consolidation is seen in the fact that the fore limbs come under the dominance of the path before the hind limbs do, excepting that in the lowest mammals the hind limbs seem never to become involved in its sphere of action. In phylogenetic development, also, this path becomes progressively more intimately consolidated with the spinal motor mechanism by shifting in the series of mammals from a dorsal to a ventral position in the spinal cord.

In *Amblystoma*, as cited in the second lecture, this growth of the associational neurones into the motor mechanism, and the functional mechanization effected thereby, is observed to occur as an expression of preneural forces of integration. It is probable, therefore, that such forces are factors in the process so long as the progressive ingrowth of the associational into the motor system continues. Habit formation, since it appears to be the behaviour counterpart of this progressive neural consolidation, may then be regarded as, in part, a function of growth in the neural mechanism. It may be perfected in a high degree without the participation of a cerebral cortex. But since the cerebral cortex represents embryonic overgrowth more or less beyond the capacity of the animal to express its immediate nervous potential, the degree to which its potential is expressed in habit must depend upon the degree of consolidation of the cortical mechanism with the motor system. The cerebral cortex of one species may accordingly have a significance very different from that in another, not simply because of different degrees of intrinsic development but because of different degrees of mechanization by ingrowth into the motor system. This is an important consideration in the interpretation of the relation of the cerebral cortex to habit formation in lower mammals*.

GROWTH AND SPECIFICITY OF FUNCTION IN NERVOUS MECHANISMS

In the first lecture it has been shown how neurones of the motor system establish the function of locomotion through the growth of axone terminals or collaterals and dendrites into particular synaptic relations (Fig. 11). In the act of growth alone, specificity of function is acquired by these elements of the mechanism: the structural relation into which they grow determines their specific function. This specificity does not have relation to any particular experience of the animal, for impulses from all sources of stimulation, external and internal, flow through the primary neural motor mechanism to reach the muscles (18). It has reference, on the other hand, only to the form of the behaviour pattern; and it is determined, in so far as our present knowledge goes,

* Lashley has observed that a large part of the cerebral cortex of the rat can be extirpated without affecting the ability of the animal to learn habits (36). This is evidence of lack of consolidation of this animal's cerebral cortex with the motor mechanism. In this respect the cerebral cortex of the rat may represent the functional grade of a considerable portion of the cerebral cortex in man; whereas in man the phylogenetically older portions of the cerebral cortex have become intimately consolidated with the neuromotor system.

exclusively by the laws of growth within the organism. In the mechanism that conditions behaviour, on the contrary, the experience of the individual, as we shall see at once, is a factor in determining the specificity of function of the constituent neurones.

The simplest type of conditioning mechanism known in the vertebrates is the Rohon-Beard cell as it occurs in *Amblystoma* (Fig. 9). A cell of this type, described in the first lecture, has its cell body within the spinal cord, has an ascending and a descending process in the cord, and another branch which passes out of the cord as a sensory fibre (13). Some of these fibres divide into two branches, one of which is distributed to muscle and the other to the skin. Such a cell is, therefore, both skin-sensory (exteroceptive) and muscle-sensory (proprioceptive). It fuses, or associates, these two modes of excitation into one mode of nerve impulse.

For the present discussion there are three things of particular importance about these most primitive cells of association: first, they establish their typical relation to skin and to muscle before the animal has any ability to respond to stimulation of the receptors, and for that reason must acquire their structural relations as elements in the mechanism of conduction through laws of growth, without activation by factors of experience in the ordinary sense; second, they give the animal no possibility to respond specifically either to external (exteroceptive) or to internal (proprioceptive) stimulation, that is to say, they lack specificity as conductors with reference to sensory modes; but, third, they have acquired differential sensitivity with reference to the quality of excitation, inasmuch as one dendritic terminal is sensitive to one mode of stimulation and another dendritic terminal is sensitive to a quite different mode of stimulation. In this development of differential sensitivity, specificity appears to be determined by interaction of the stimulating agent (skin or muscle) and the processes of growth in the dendritic terminal.

How growth and excitation can interact to determine specificity in sensitivity as they obviously do in the Rohon-Beard cell is not known; but the chronaxie of a nerve is known to be correlated with that of the muscle it innervates. There is also an approximate isochronism between the nerve and the muscle*. Chronaxie would

* "It has been found that irritable tissues are to be distinguished from each other not by their threshold of excitation, but by the *duration* of the exciting stimulus (galvanic). In general the sluggish tissues require a stimulus of long duration, while a 'rapid' tissue (which may have the same threshold for galvanic currents) is stimulable by briefer currents. For routine comparison it is essential to use a strength of current somewhat above the threshold in order to guard against adventitious alterations in tissue resistance (due to moisture, etc.). It is usual to employ a current of double the threshold intensity, and the minimal effective duration of such a current is referred to as the 'chronaxie' (Lapicque) or 'excitation time' (Lucas)." Fulton (24), p. 78 *et seq.*

"...while very few attempts have been made to apply this illuminating hypothesis to the physiology of the central nervous system, the recent experiments of Bremer and his collaborators have made it evident that the chronaxial conception will undoubtedly throw much light upon the problems of central nervous physiology. According to Lapicque, the power of exciting a tissue rests upon the duration of the exciting stimulus whether this be an induction shock, a condenser discharge, or the action current of a nerve. In the case of nerve and muscle the Lapicque theory postulates that the action current at the motor end-plate is of approximately the same duration as that of the local excitatory process necessary for stimulation of the muscle. For simplicity the 'characteristics' of the excitatory process of a given tissue have been measured artificially by condenser discharges and the least duration of stimulus (of double the threshold seem, therefore, to be an important factor in the sensitivity of muscle to specific excitation from its nerve. It is not unreasonable, therefore, to infer that the same principle is involved in the excitation of nerve by muscle; for, in the case of the Rohon-Beard cells, the muscle cells may impress their own chronaxie upon the muscle-sensory endings, and, if chronaxie operates in the excitation of nerve by muscle, it may operate also in the excitation of nerve by sensory cells. Furthermore, if chronaxie operates in determining specificity of sensitivity in nerve endings with reference to end organs, it may operate also in the excitation of one nerve cell by another. This, of course, is a far-reaching inference, for it hypothecates chronaxie and isochronism at every synapse in the nervous system.

Following upon these inferences it may be noted that, in higher vertebrates, the differential specificity of sensitivity, such as appears in the Rohon-Beard cell, is projected into the neurones as a whole, in the sense that

intensity) necessary for excitation has been termed the 'chronaxie.' Lapicque finds that the chronaxie of a motor nerve is approximately that of the muscle which it innervates. This condition is referred to as 'isochronism.' Now certain drugs have the power of acting preferentially on the time-relations of one or other of the two neuromuscular constituents, *i.e.* upon the nerve or the muscle. When a nerve-muscle preparation is poisoned, *e.g.* by strychnine which acts solely on the nerve, it is found that the chronaxie of the nerve diminishes, and when it has diminished to about half its normal value the nerve becomes then unable to excite its muscle. It may so happen that the muscle itself may have been acted upon by drugs which reduces its chronaxie (*e.g.* nicotine) and a nerve block produced in this way. When strychnine is applied to such a preparation it will restore conductivity owing to the fact that it reduces the chronaxie of the nerve in the same direction as nicotine has reduced it in the muscle." Fulton (24), p. 463 *et seq.*

some of the afferent neurones become specific for excitations in the skin and others become specific for excitations in muscle; and this specificity of function goes over even into the neurones of the second or even higher order in mammals. But under these conditions there must be somewhere in the circuits of the conditioning mechanism neurones of the same functional value as the Rohon-Beard cells, in that they fuse (associate) excitations of different modes into a single potential of conduction. Also, wherever such cells occur, it is reasonable to infer that their dendritic terminals acquire specificity in sensitivity, as in the case of the Rohon-Beard cell, by interaction between the processes of growth and excitation.

DIFFERENTIAL SENSITIVITY IN THE CEREBRAL CORTEX

The Rohon-Beard cell may be taken, then, as the prototype of the associational or conditioning mechanism in the nervous system of vertebrates. As such it has important bearing upon physiological and psychological interpretations because, representing as it does the mechanism of association in the simplest terms, it is definitely known with regard to its development, structure and function. It is known to associate stimuli of different modes through different dendrites and it is known to extend its dendritic system enormously by growth after it begins to function as a conductor.

At the other extreme of association-cell types are the pyramidal cells of the cerebral cortex, or the Purkinje cells of the cerebellum. But one of these cells, as compared with its prototype, is almost as a man compared with an amoeba. Nevertheless, a pyramidal cell or a Purkinje cell may be conceived as functionally a Rohon-Beard cell distantly removed from the receptor field and enormously elaborated in its system of dendritic terminals and in its multiplicity of modes of excitation through the terminals of axones of nerve cells from various sources. That they, also, grow extensively after they begin to function seems unquestionable.

The degree of structural development of the pyramidal cells in man when the receptors begin to function is not known; but it is known that Betz cells are highly differentiated in a human fetus of eighteen weeks (6). At that age it would seem, upon anatomical grounds, that a considerable portion of the human cerebral cortex is capable of functioning as a conducting mechanism. Its source of excitation, however, must be very limited as compared with adult conditions. Only a muscle-sensory (proprioceptive) and, possibly, deep pressure sensitivity can be regarded as active fields of stimulation at that time. Under these modes of excitation the pyramidal cells and other nerve cells then functional in the cortex may be regarded as determining specificity of sensitivity in their growing dendrites, according to the mode of stimulation which impinges upon them through axone terminals from different sources in the same manner that Rohon-Beard cells acquire differential specificity through excitation by muscle on the one hand and by skin on the other. As development proceeds, new dendrites grow out from the cortical cells simultaneously with the ingrowth of new axone terminals; and so long as such growth continues the cortical cells, according to this hypothesis, progressively determine a multiplicity of grades or modes of sensitivity. This process of differentiation of sensitivity may be thought of as progressing in the main along the axis of the pyramidal cell at right angles to the plane of cortical lamination; but not exclusively, for large dendrites grow also in the plane of lamination, particularly in the deeper laminae, which, according to Bolton's interpretation of localization of cortical function, have to do with more primitive, general or instinctive behaviour, as opposed to the more refined mental and social adaptation (5).

According to this interpretation, particular regions of the cerebral cortex have to do pre-eminently with the more subtle adaptations in behaviour not because of their topographical position in the brain but because of the period in the life of the individual when they begin to participate in the conditioning of behaviour. Cells or parts of cells that begin to function in this process early are dominated by the earlier and more primitive functions because their specific sensitivity was established in relation to the neurones that served those functions, whereas cells or parts of cells that rise into function later are only indirectly influenced by primitive functions and are pre-eminently dominated by situations or adaptations of later life. This interpretation is obviously in accordance with Bolton's theory of cortical localization of function according to the pattern of lamination (5).

How long growth of dendritic terminals continues in the brain of mammals is not known, and our available cytological technique is not adequate to determine. But the inference seems reasonable that such growth continues as long as the cortical cell bodies increase in size, and there is considerable evidence to show that there are cortical cells that increase in size until the brain acquires its full weight and gross dimensions. But it is not impossible that dendritic terminals of very small microscopic dimensions continue to grow for even a longer period and that there is compensation for their increase by decrease of other tissue.

But be that as it may, cells in the cerebral cortex of mammals certainly grow after they begin to function as conductors. As in *Amblystoma*, so in mammals, the nerve cell in the most embryonic parts of the brain, which is the cerebral cortex, must be thought of, not simply as a conductor, but also as an element in a system of preneural forces that are constantly reacting upon one another. Its reaction to its environment through its inherent potentiality of growth must play a part in the conditioning of behaviour in the mammal as does similar reaction of the primitive association cell in *Amblystoma*. One phase of this activity must be the development constantly of new specificities of sensitivity through growing dendritic terminals in relation to the different modes of stimulation. It may, indeed, be that such a progressive differentiation of sensitivity essentially constitutes the conditioning process.

STRUCTURAL COUNTERPARTS OF BEHAVIOUR

In the first lecture aquatic locomotion and the earliest limb movement in *Amblystoma* were explained upon the basis of the organization of the nervous system (18). The neural mechanism in this case is potentially the behaviour, in so far as the form of the pattern is concerned (Figs. 9, 10, 11). The mechanism we described may therefore be regarded as the structural counterpart of the form of the behaviour pattern.

It is possible, also, that conditioning processes are registered in structural counterparts in the sense that neural mechanisms acquire functional specificity with reference to the experience. In the counterpart of the form of the pattern, as already explained, the specificity of function is fixed by the relations into which the elements grow. In the counterpart of experience, on the other hand, specificity of function is established by interaction of growth and excitation, that is to say, the excitation fixes upon the growing terminals of neurones its own mode of activation. In the conditioning mechanism in general, as in the case of the Rohon-Beard cell, according to this hypothesis, laws of growth determine the structural relation of conductors, but their specific sensitivity is fixed by the mode of excitation.

In the motor mechanism of *Amblystoma* we see structural counterparts of attitudes which are released into action of definite form in appropriate situations. It is possible that in the conditioning mechanisms, also, situations organize themselves into definite structural counterparts through the interaction of growth and excitation.

"GESTALT" IN THE CONDITIONING MECHANISM*

This conception of growth as a component of the function of the nerve cell in conditioning behaviour is

* Ogden has translated "Gestalt" as "configuration": "But what can be said of those experiences which put us in touch with the outer world? How are the perceptions of an infant constituted? We find the new-born child capable of movement whenever external stimuli come in contact with his senses; that is, whenever the equilibrium

consistent with the fact that, normally, the behaviour pattern is a unity from first to last in the career of the

of his condition is disturbed. For instance, a bright object appears in the field of vision and the eyes move; a contact is made with a certain place on the hand and the fingers close, etc. In every case a state of rest is interrupted; into the already existing world wherein the child was at rest a new factor has been introduced which disturbs his quiescence. If we wish to reconstruct the phenomenal counterpart of this objective behaviour we must consider the child's state as a *whole*. Consequently, we ought not to say that the child sees a luminous point; but rather that the child sees a *luminous point upon an indifferent background*; or, in the case of touch, that pressure is felt upon the hand, otherwise untouched. Generally stated, *from an unlimited and ill-defined background there has arisen a limited and somewhat definite phenomenon, a quality*. Whether or not the background existed phenomenally even before the new factor emerged from it, will be discussed later. Here it is sufficient to note that when a quality appears, the 'indifferent' ground must also be considered as more or less 'uniform.' We are presupposing that before the appearance of the stimulus the child was at rest, and not moving. Inferring phenomena of experience from behaviour, an undifferentiated phenomenon must correspond to the absolutely undifferentiated behaviour of quiescence. The reader should not forget that we are speaking of the earliest beginnings of consciousness; and that it is the very first experience of the child that we are attempting to characterize. Our characterization is, then, this: That the first phenomena are *qualities upon a ground*. Introducing at this point a new concept, they are the simplest *mental configurations*. The phenomenal appearance in consciousness divides itself into a given quality, and a ground upon which the quality appears—a level from which it emerges. It is, however, a part of the nature of a quality that it should lie upon a ground, or, as we may also say, that it should rise above a level. Such a co-existence of phenomena in which each member 'carries every other,' and in which each member possesses its peculiarity only by virtue of, and in connection with, all the others, we shall henceforth call a *configuration*. According to this view, the most primitive phenomena are figural; as examples, the luminous point set off from a uniform background; something cold at a place upon the skin set off from the usual temperature of the rest of the skin; the too cold or too warm milk in contrast with the temperature level of the mouth-cavity. We attribute configurations, also, to such individual, for cortical cells, beginning their function with the beginning of experience, grow as experience progresses till all of the essential behaviour and conditioning processes are registered in them. Every pyramidal cell as a growing unit may be conceived as blending, so to speak, the experience of the individual from the beginning to the end of stimulation and response, after the manner that the Rohon-Beard cell blends exteroceptive and proprioceptive excitations. As a result of this, although the behaviour at any moment may be dominated by some particular phase of experience, it cannot be utterly disconnected from any part of the whole. Only with the retrogressive changes of senescence or with arrested development in pathological cases does experience cease to register in a progressive manner. Nevertheless, even under such conditions, the ex-

reactions as the rejection of milk when it is not of the right temperature; thus milk in the mouth may lead either to an 'adequate' or to an 'inadequate' configuration.

To many this view of the constitution of the most primitive phenomena will appear very odd indeed: for it assumes that a certain order dominates experience from the beginning, whereas we would be in much better agreement with current views if we were to assume that order comes only as a result of experience—a theory which has given rise to the view that the consciousness of the new-born infant is nothing but a confused mass of separate *sensations*, some of which are present earlier than others, because of the earlier maturation of their appropriate brain-centres. Upon the basis of such a theory the sense of vision would seem to supply the child with a chaotic mass of achromatic and chromatic impressions, like the colours upon a painter's palette, from which experience would proceed to choose the ones that are requisite to constitute his perceptual world. And this doctrine is founded upon one of the fundamental presuppositions with which psychology has long worked; namely, that single mental units called sensations are aroused in a simple manner by stimulation, and from them every other kind of experience is derived by a process of association." Koffka (33), p. 130 *et seq.*

perience that is actually operative in conditioning the behaviour is, according to this conception, a continuum by reason of the growth potentiality of the neurone. Every new adjustment arises out of the total of the individual's experience as registered in structural counterparts by means of the correlation of growth and excitation in the developing neurones. Or in terms of "Gestalt" the particular adjustment is related to the total experience as a "quality upon a ground."

GROWTH AS THE CREATIVE FUNCTION OF THE NERVOUS SYSTEM

Conduction, however complex the organization may be, cannot fully account for the rôle of the nervous system in behaviour. The conventional figure of the telephone with its switchboards to illustrate how the nervous system works is utterly inadequate, unless the inventor and the operator of the telephone be included in the figure. In the nervous system the growth potential is at once the inventor and the operator.

In the early development of *Amblystoma*, as already described, there is a great deal more central nervous organization than can express itself through the effectors, and certain elements in this overgrowth can be recognized as elements in the structural counterparts of future behaviour forms. It has been noted also that the same kind of neural overgrowth involves the conditioning mechanism. With application of the conception of structural counterparts of experience to the conditioning mechanism, it is conceivable that in this continuously growing embryonic matrix, neural mechanisms of specific behaviour value may arise which have no possibility of immediate expression. Such mechanisms must be organized out of elements of experience, but they are essentially new creations as regards their real identity. They represent such factors of behaviour as attitudes which can come to an issue only in the more or less distant future, but which, at the appropriate time, issue in decisive, essentially predetermined action. This predetermination may be regarded, in mechanistic terms, as an act of will. It may arise within the constantly expanding conditioning mechanism by a process of individuation in the same manner as the limb reflex emerges within an expanding total behaviour pattern. In the latter case the mechanism is in process of creating a definite end-result for a long period before the end-result is attained. So, also, in the conditioning mechanism it is conceivable that similar creative acts of growth may be in process of eventuation in behaviour far in the future. Growth, accordingly, may be conceived as the creative function of the nervous system, not only with regard to the form of the behaviour pattern, but also with regard to its control. The creative component of thought, upon this hypothesis, is growth.

THE MEASURE OF THE INDIVIDUAL

If, then, it is conceded that growth is one of the means by which the nervous system performs its function in behaviour, it must be granted, contrary to the dogma of certain behaviourists, that man is more than the sum of his reflexes, instincts and immediate reactions of all sorts. He is all these plus his creative potential for the future. Even the embryo of *Amblystoma* is, mechanistically considered, more than the sum of its reflexes or immediate behaviour possibilities. The real measure of the individual, accordingly, whether lower animal or man, must include the element of growth as a creative power. Man is, indeed, a mechanism, but he is a mechanism which, within his limitations of life, sensitivity and growth, is creating and operating himself.

LITERATURE CITED

(1) Äriens Kappers, C. U. (1917). Further contributions on neurobiotaxis. IX An attempt to compare the phenomena of neurobiotaxis with other phenomena of taxis and tropism. The dynamic polarization of the neurone. *Journ. Comp. Neur.* **27**, no. 3.

(2) Baker, R. C. (1927). The early development of the ventral part of the neural plate of *Amblystoma*. *Journ. Comp. Neur.* **44**, no. 1.

(3) Bodine, J. H. and Orr, P. R. (1925). Respiratory metabolism. *Biol. Bull.* **48**, no. 1.

(4) Bok, S. T. (1915). Die Entwicklung der Hirnnerven und ihrer Zentralen Bahnen. Die Stimulogene Fibrillation. *Folia Neurobiologica*, **9**.

(5) Bolton, J. S. (1910, 1911). A contribution to the localization of cerebral function, based on the clinico-pathological study of mental diseases. *Brain*, **33**.

(6) Bolton, J. S. and Moyes, J. M. (1912). The cyto-architecture of the cerebral cortex of a human foetus of eighteen weeks. *Brain*, **35**, part 1.

(7) Child, C. M. (1915). *Senescence and rejuvenescence*. Chicago.

(8) —— (1915). *Individuality in organisms*. Chicago.

(9) —— (1921). *The origin and development of the nervous system from a physiological viewpoint*. Chicago.

(10) —— (1924). *Physiological foundations of behaviour*. New York.

(11) Coghill, G. E. (1902). The cranial nerves of *Amblystoma tigrinum*. *Journ. Comp. Neur.* **12**, no. 2.

(12) —— (1906). Cranial nerves of *Triton taeniatus*. *Journ. Comp. Neur.* **16**, no. 4.

(12*a*) —— (1908). The development of the swimming movement in Amphibian embryos. *Anat. Rec.* **2**, no. 4, p. 148.

(12*b*) —— (1909). The reaction to tactile stimuli and the development of the swimming movement in embryos of *Diemyctylus torosus*, Eschscholtz. *Journ. Comp. Neur.* **19**, no. 1.

(12*c*) —— (1913). The primary ventral roots and somatic motor column of *Amblystoma*. *Journ. Comp. Neur.* **23**, no. 2.

(13) —— (1914). Correlated anatomical and physiological studies of the growth of the nervous system of Amphibia. I. The afferent system of the trunk of *Amblystoma*. *Journ. Comp. Neur.* **24**, no. 2.

(14) COGHILL, G. E. (1916). II. The afferent system of the head of *Amblystoma*. *Journ. Comp. Neur.* **26**, no. 3.

(15) —— (1924). III. The floor plate of *Amblystoma*. *Journ. Comp. Neur.* **37**, no. 1.

(16) —— (1924). IV. Rates of proliferation and differentiation in the central nervous system of *Amblystoma*. *Journ. Comp. Neur.* **37**, no. 1.

(17) —— (1926). V. The growth of the pattern of the motor mechanism of *Amblystoma punctatum*. *Journ. Comp. Neur.* **40**, no. 1.

(18) —— (1926). VI. The mechanism of integration in *Amblystoma punctatum*. *Journ. Comp. Neur.* **41**, no. 1.

(19) —— (1926). VII. The growth of the pattern of the association mechanism of the rhombencephalon and spinal cord of *Amblystoma punctatum*. *Journ. Comp. Neur.* **42**, no. 1.

(20) —— (1928). VIII. The development of the pattern of differentiation in the cerebrum of *Amblystoma punctatum*. *Journ. Comp. Neur.* **45**, no. 1.

(21) DETWILER, S. R. (1923). Experiments on the reversal of the spinal cord, in *Amblystoma* embryos, at the level of the anterior limb. *Journ. Exp. Zool.* **38**, no. 2.

(22) —— (1927). Experimental studies on Mauthner's cell in *Amblystoma*. *Journ. Exp. Zool.* **48**, no. 1.

(23) ELLIS, R. S. (1919). A preliminary quantitative study of the Purkinje cells in normal, subnormal, and senescent human cerebella, with some notes on functional localization. *Journ. Comp. Neur.* **30**, no. 2.

(24) FULTON, J. F. (1926). *Muscular contraction and the reflex control of movement*. Baltimore.

(25) HARRISON, R. G. (1910). The outgrowth of the nerve fibre as a mode of protoplasmic movement. *Journ. Exp. Zool.* **9**, no. 4.

(26) HERRICK, C. J. (1924). *Neurological foundations of animal behaviour*. New York.

(27) —— (1926). *Brains of rats and men: a survey of the origin and biological significance of the cerebral cortex*. Chicago.

(28) HERRICK, C. J. and COGHILL, G. E. (1915). The development of reflex mechanisms in *Amblystoma*. *Journ. Comp. Neur.* **25**, no. 1.

(29) HOOKER, D. (1917). The effect of reversal of a portion of the spinal cord at the stage of closed neural folds on the healing of the cord wounds, on the polarity of the elements of the cord and on the behaviour of frog embryos. *Journ. Comp. Neur.* **27**. no. 4.

(30) INGVAR, SVEN (1920). Reaction of cells to the galvanic current in tissue cultures. *Proc. Amer. Soc. Exp. Biol. and Med.* **17**.

(31) INUKAI, T. (1928). On the loss of Purkinje cells, with advancing age, from the cerebellar cortex of the albino rat. *Journ. Comp. Neur.* **45**, no. 1.

(32) KINGSBURY, B. F. (1926). The so-called law of antero-posterior development. *Anat. Rec.* **33**, no. 2.

(33) KOFFKA, KURT (1924). *The growth of the mind. An introduction to child psychology.* Translated by R. M. Ogden. New York.

(34) LANDACRE, F. L. (1921). The fate of the neural crest in the head of the urodeles. *Journ. Comp. Neur.* **33**, no. 1.

(35) LANGWORTHY, O. R. (1927). Histological development of cerebral motor areas in young kittens correlated with their physiological reaction to electrical stimulation. *Contr. to Embryol.* no. 104; *Carneg. Inst. Wash. Pub.* no. 380.

(36) LASHLEY, K. S. (1920). Studies of cerebral function in learning. *Psychobiol.* **2**, no. 1.

(37) MAGNUS, R. (1925). Animal posture. Croonian lecture. *Proc. Roy. Soc.* (B), 98 *B*.

(38) MATHEWS, A. P. (1903). Electrical polarity in the hydroids. *Amer. Journ. Phys.* **8**, no. 4.

(39) PAVLOV, I. P. (1927). *Conditional reflexes: an investigation of the physiological activity of the cerebral cortex.* Translated and edited by G. V. Anrep. Oxford.

(40) RAMÓN Y CAJAL, S. (1904). *Textura del sistema nervioso del hombre y de los vertebrados*, **2**, pp. 1150–1152. Madrid.

(41) SHERRINGTON, C. S. (1906). *The integrative function of the nervous system.* New York.

(42) TRACY, H. C. (1926). The development of motility and behaviour reactions in the toadfish (*Opsanus tau*). *Journ. Comp. Neur.* **40**, no. 2.

CAMBRIDGE: PRINTED BY
W. LEWIS, M.A.
AT THE UNIVERSITY PRESS